Penguin Education

Race, Culture and Intelligence

Edited by Ken Richardson and David Spears
Associate Editor: Martin Richards

Race, Culture and Intelligence

Edited by Ken Richardson
and David Spears
Associate Editor: Martin Richards

Penguin Books

Penguin Books Ltd, Harmondsworth,
Middlesex, England
Penguin Books Inc, 7110 Ambassador Road,
Baltimore, Md 21207, USA
Penguin Books Australia Ltd,
Ringwood, Victoria, Australia

First published 1972

Made and printed in Great Britain by
C. Nicholls & Company Ltd
Set in Monotype Plantin

Contents

Editorial Foreword

This collection of essays about intelligence stems from the revived nature–nurture controversy about the origins of mental abilities, led notably by Arthur Jensen whose article in 1969 created a furore in the USA, and more lately by H. J. Eysenck in Britain. Their position, ostensibly vindicated by dispassionate appeal to scientific evidence, attributes the persistent IQ differences between US blacks and whites predominantly to differences in genetic constitution.

The heated exchanges which followed the publication of Jensen's article were felt on both sides of the Atlantic. One culmination of this was the organization of a public debate by a study group of the Cambridge Society for Social Responsibility in Science, at which Arthur Jensen defended his views. But the wider debate has continued with demands for fresh research on the one side, and appeals for composure on the other, while the layman has become exposed to arguments which formerly had been restricted to academic circles. This is what brought together a group of educationists and scientists at the Open University, who discussed the social implications of these developments and who, equipped with concurring views from other centres, decided to assemble a popular book, intended to explain the arguments to the lay-reader. It was in this atmosphere of interdisciplinary concern for the human aspects of science that the foundations were laid and the proposed design and content were evolved. Similar intentions had been latent in moves by the Cambridge group, and their subsequent cooperation has enhanced the production of this book which combines contributions from psychologists, sociologists, educationists and biologists who have been engaged in relevant work.

In planning this book we have attempted to step back from the debate itself and look at the concepts which underlie it. This involves a close examination of the key ideas – intelligence, race, heredity and environment – as well as following some of the implications of the evidence for our complex, heterogeneous society. The authors have attempted to disentangle this evidence from the fine mesh of ideology, value-judgement, assumption and misconception, where it is so often found. The issues are complex but as far as possible are presented in ways that the non-specialist can follow.

After an opening chapter by Liam Hudson which sets the stage and provides the historical background, the book divides into three parts. The first is psychological and deals with the nature of intelligence, its development and relationship to school progress. The second part is the domain of the biologists, who discuss the genetics of IQ and intelligence, and the interpretation of race differences in these capacities. Then attention is turned to development and a consideration of environmental influences on brain growth. In the final part the scope is broadened to look at the social world, both as the context for the development of intelligence, and as the context for the debate about race differences. The concluding chapter draws together the major threads of the argument and discusses some of the educational implications, particularly those for compensatory programmes. We do not offer any final answers, indeed, a major aim is to show that questions have been often posed in ways that cannot yield them, but we hope to offer a clear perspective for the whole problem and a starting point from which we may proceed less divisively.

The production of this book has been in many ways a team effort and represents the participation of many persons not represented in these pages. We would particularly like to thank Brian Lewis and Mike Macdonald-Ross who played major roles in launching this project, and John Stewart for sustained interest and practical support. We are grateful to John Dobbing of Manchester University and the staff of the Psychology Department of North East London Polytechnic for fruitful collaboration. Many others, especially Tom Blair and John Simpson, commented on the initial abstracts and Brian Tiplady, Jeff Haywood, Arun Sinha, Javad Hashteroudian and Robin Harding have afforded us continuous support and encouragement. Finally, we would like to thank those original members of the CSSRS Study Group (Judy Bernal, Susan and Ken Kaye, Elena Leiven and Caroline Hall) who have not written for this book but who helped to form one of the strands of its conception.

Much advice has been incorporated from these sources but of course, the editors take responsibility for the book in its final form.

The Editors

The Context of the Debate

Liam Hudson

Liam Hudson has been Professor of Educational Sciences at Edinburgh University since 1968. Previously, he worked for eleven years at Cambridge, first as a member of the Psychological Laboratory, and from 1966 as a Fellow of King's College. His published works include *Contrary Imaginations, Frames of Mind* and *The Ecology of Human Intelligence*.

After a period of quiet, one that lasted from the end of the Second World War until the late 1960s, the issue of race and intelligence is once again in the news. The debate is inconclusive and at times bitter – not merely men's jobs, but their lives, have been threatened for the views they have expressed. For this reason, it is an unusually difficult debate for the layman to disentangle. The arguments employed are drawn from two spheres of human life that we have learnt to consider as quite separate: science and politics. Participants seem to slip continually from one sphere to the other; at one moment playing the role of the dedicated scientist, at the next that of the politician – and technical arguments, on both sides, are used to advance points of view that are essentially ideological. At one extreme, there are those who claim that 'the truth must out'. They then go on to say, with a reluctance that does not quite ring true, 'and, as it happens, research proves black men to be inferior to white'. At the other extreme, there are those who argue that such research is morally outrageous and should be suppressed. Those who use scientific evidence to reach conclusions about the inferiority of the blacks accuse their opponents of conspiring to suppress the truth. Their opponents retort that the self-consciously scientific are mediocre scientists who have been muddled by their own statistics. To make matters worse, the issue has been taken up by the mass media, and even the most abstemious now find themselves encouraged to act a part on the public's behalf, trying to undermine the opposition's case, and to attract as many recruits as possible to their own side. Issues of logic are clouded by all the tricks of the persuader's art. Experts establish their own expertise, whilst casting doubt on the legitimacy or motives of the opposition; red herrings are proffered, veiled insults are bandied, and there is moral posturing on all sides.

The truth of the matter is that most of the utterances made by scientists about race and intelligence are devoid either of scientific validity or educational significance. But to acknowledge this is not to dismiss the issue out of hand. Not all who argue for the influence of human genetics are racists; not all who argue from the environmental point of view are mindless egalitarians. There exists a middle ground, and in the United Kingdom at least, it is still possible that questions of race and intelligence can be pursued as possessing a certain academic interest in their own right. The evidence collected so far strikes most commentators as circumstantial, to say the least. But the possibility remains that racial and social groups do differ in their intellectual potentialities, and that some clearer idea of the causes underlying these differences may eventually be reached. The task, though, is one that neither the geneticist nor the social scientist can reasonably hope to master alone. It may well be that the issues of race and intelligence will lead, in the long run, to a collaboration; and that this collaboration, in its turn, may lead to the resolution of problems of greater intrinsic importance than why, on average, black children should differ from white by a certain number of points in IQ. Further, and perhaps more valuable still, we may learn more about the subtle ways in which an individual's personal and political loyalties enter into what he believes is objective research.

The first step in making sense of this long-standing dispute is to establish it in its historical and political context. In this way, it becomes easier to identify the more naive and extreme positions adopted, and to clear a no-man's land for more level-headed debate.

The belief that some races are genetically inferior to others was a pronounced element of Victorian thought, and the psychologist Francis Galton, in his remarkable book *Hereditary Genius*, certainly stated it in the most uncompromising terms. 'It is,' he said, 'in the most unqualified manner that I object to pretensions of natural equality.' He goes on to say that 'The mistakes that the Negroes made in their own matters were so childish, stupid and simpleton-like, as frequently to make me ashamed of my own species.' Galton was the founding father of intelligence testing; and many of his followers, the mathematician Karl Pearson, for example, shared his beliefs. Two schools of thought gradually took shape – the hereditary and the environmental – and the differences between them flared, in the 1930s, into a dispute as virulent as it was unedifying. At that time, the hereditary school drew their support, largely but not exclusively, from those trained in the biological and physical sciences. Their opponents drew theirs, again largely but not exclusively, from those trained in the social sciences. The argument split psychology down the middle.

In restrospect, it is hard to see this dispute among academics as anything other than the reflection within psychology of stresses then mounting in society at large. It can scarcely be a coincidence that the debate was reaching its crescendo at exactly that historical moment when the Nazis were using the hereditary view as a justification for genocide – their attempt to exterminate the Jews. Once the War was over, and the atrocities of the death camps became public knowledge, the hereditary view – with certain geographical exceptions – became taboo. In South Africa, hereditary arguments were still used as the justification for apartheid, and in the southern states of America they were used to justify the social repression of the blacks. Elsewhere, such ideas were banished from public debate. Now they have resurfaced; and it is important, before examining their merits, that we should try first to grasp the reasons for the present change of mood.

Since the Second World War, the possibility that most black people might be born with lower intellectual potentialities than most white people, has been glossed over in a series of comfortable, liberal evasions. These now seem scarcely more satisfying than the prejudices they replaced. Where Nazis believed that observed differences were caused by 'blood', post-War liberals assumed, without evidence, that such differences were either illusory, or, if real, that they were to be explained by differences of environment and opportunity. And in the propagation of these liberal, egalitarian beliefs, the rapid growth of university courses in social science seems to have had an important part to play. As recently as the mid-1960s, students were emerging from such courses in British universities believing that human beings adopt, putty-like, whatever shape environmental forces thrust them into; and that, in particular, any utterance about differences between racial groups was either false or meaningless. Differences between Einstein and Hitler, between Picasso and the man who sweeps the factory floor in Detroit or Dagenham – all these, it was implied if not formally stated, were self-evidently accountable in terms of the historical and social circumstances in which each individual grew up.

Doctrinaire environmentalism of this kind carried with it a halo of further beliefs; together they constitute what might crudely be described as the 'libertarian ticket'. The thrust of undergraduate teaching in the social sciences was to undermine our self-centredness, our 'ethnocentricity'. Students were taught that human nature and human society were in no sense immovably fixed. In the various corners of the earth, anthropologists had recorded the existence of societies totally different from our own, but nonetheless viable for that. Yet, paradoxically, many university teachers in the social sciences seem to have drawn from this anthropological evidence the conclusion not merely that all human conventions are 'relative', but that they are optional, in the sense that rational people can dispense with them altogether. As a consequence, all arbitrary-seeming regularities in our own society were brought under attack: differences between men and women, the conventions of sexual morality, and patterns of parental and social authority. Most interestingly perhaps, from the point of view of the present debate, the authority of the university teacher over his own students was itself brought seriously into question.

In each of these respects, social scientists may have done no more than act as an *avant-garde*, reflecting a more general reaction from the rigid demarcations and hypocrisies of Victorian society. It should scarcely surprise us, on the other hand, that such teaching should provoke a reaction – the 'authoritarian backlash'. Nor should it surprise us that the resurgence of interest in hereditary arguments about race should have been triggered in societies experiencing mounting racial tension, and the distressing side-effects of a movement towards student power. The writings of the hereditary school are uniformly tough and authoritarian in tone: they assert the primacy of science above all other considerations; they stress technical expertise; they employ, behind a dazzling display of statistics, concepts that are evocative but simple; and they endorse the virtues of an orderly and conventional society. The barbs, in the works of Professors Jensen and Eysenck, are directed not against black people, nor any other racial minority, but against the iniquities of their sociological colleagues. It follows that the debate about race and intelligence only begins to make sense when it is seen as one internal to academic life; between two groups of men who differ in personality, in academic background, and in political and social allegiance.

So much for the broad context of the present debate. There remains, still, some scene-setting to do within psychology itself. For much of the debate centres on a single technical device – the IQ test – and, as a result, it suffers from a curious sense of artificiality. The explanation is an unhappy one, and is to be found in the snobbery that exists among psychologists themselves. For the field of intelligence, over the last thirty years, has enjoyed very low prestige indeed. Research funds have been sparse, able students have been encouraged to specialize elsewhere, and undergraduate teaching departments have treated the subject as unworthy of serious attention. As a result, the study of human intelligence, an area that should be among the most exciting that science can offer, has stagnated to an alarming degree. It is not merely that there is disagreement about details of technique and niceties of interpretation; there exists disagreement of a quite fundamental kind about how research findings can be brought to bear on issues of educational and social importance. In psychology, as elsewhere in science, the 'facts' do not speak for themselves; and nothing is easier in psychology than generating large quantities of facts that look vaguely interesting or important, but which in truth have no logical implication at all.

For reasons that are still unclear, the use of IQ tests has in fact taken on many of the qualities of a mystic rite. The IQ has come to be seen as a measurement that not merely summarizes the individual's capacity to perform certain tasks, but one which, in some unspecified way, puts a number to his essential worth. To have a low IQ is seen as the equivalent of having low caste. That such a system of fantasy should surround a simple, useful, but prosaic mental testing procedure is odd; but its implications are of the greatest importance. For, in insidious ways, the assertion that a black man has a lower IQ than a white man becomes tinged – in the minds of psychologist and layman alike – with implications about those individuals' basic value. As a consequence, the use of technicalities, far from rescuing us from the realm of morals, makes the discussion of intelligence more slippery than ever before.

Even when the IQ test is seen for what it is, we still face difficulties of logic: the question of what our evidence can sensibly be used to prove. We may grasp that the IQ is not some simple, quasi-physical attribute (like a car engine's horse power), yet still fail to realize how tight are the logical restrictions that surround us.

At the moment, and probably in principle, it is impossible to design an experiment that separates out in any neat and tidy way, the influence of hereditary and environmental factors on a human being's ability to do IQ tests. But let us grant, for the sake of argument, that Peter is better than Paul at all the tests of reasoning that we can devise. Let us also grant that Peter's superiority is largely genetic in origin. What can we conclude? We certainly cannot conclude that Peter will perform better than Paul at any job involving brainwork. Far from it. Peter may be clever but lazy, tactless, ineffectual; Paul may be slower on the uptake, but persevering, sensitive with people, and a good leader. These personal qualities may be more important in determining how Peter and Paul work than any purely intellectual ones. Does our evidence tell us, in the case either of Peter or of Paul, where their intellectual limits lie? Again, no. If their ways of life are rewarding, both Peter and Paul may think with increasing penetration and effectiveness as the years go by; at the age of forty, they may be resolving problems which, twenty years earlier, neither could have entertained. Can we conclude, even if we do not know the limits of their powers, that in purely intellectual matters Peter will always be a step ahead of Paul? Again, no. Intellectual skills are acquired by opportunity; and Peter's laziness, say, may preclude him from opportunities that Paul enjoys. At its simplest, Paul may reach university, and Peter may drop out. Quite rapidly, Paul's intellectual capabilities may have outrun Peter's.

The moral is simple: in practice, it is quite pointless to abstract ideas about intelligence from the personal and social considerations that determine the use to which our brains are put. Even greater difficulties of interpretation arise when our evidence concerns, not individuals, but average differences between populations; and they are redoubled, yet again, in the case of blacks and whites especially, when issues of social prejudice are at stake.

Our aim – in research as in teaching – is to discover what constraints limit the growth of an individual's full intellectual powers. Yet even if, for the sake of argument, we were to grant the most extreme possibility – that *all* black children are born less intelligent than *all* white children, a wildly unlikely state of affairs – we are still little the wiser. Dimly, we may feel that something of educational importance is at stake; yet when the proposition is examined in detail, its practical implications trickle away. Such a proposition does not tell us, for example, whether black children and white should be taught separately or together; it gives no clues as to how each child can be lured into the use of his brains, or where his limits may lie; and it tells us nothing about how the country's educational resources should best be spent. In as much as such questions are dependent on factual evidence, the evidence is of a kind we do not yet know how to collect; in as much as they are moral and political, they lie outside the boundaries of conventional science altogether.

From Psychology

In this section we take a look at the concept of intelligence and its measurement through IQ tests.

John Radford and Andrew Burton examine three major traditions within psychology which have taken different approaches to the question of intelligence. The first, which they call the 'London line', stems from the work of Darwin's cousin Galton, and provides the intellectual heritage for Eysenck and the whole modern school of psychometrics – the enterprise concerned with the measurement of mental abilities and personality. The authors argue that this whole tradition is shaped by its taking physics as its model – intelligence becomes a 'thing' – and its initial assumption that individual differences in intelligence stem from hereditary factors. One of the most important consequences of this approach is the idea that a person's intelligence is a stable quantity of something we can measure with a simple test. Discussion then turns to the creativity tradition which, the authors argue, was an attempt to add notions of originality, inventiveness and imagination to the cold logical idea of intelligence. Progressive schools were intended to foster creativity but their results seem uncertain and, in fact, merely beg questions about the aims and methods of education. What kinds of cognitive capacities should we try and foster in children? In the final part of the chapter Piaget's work is discussed. Here biological, not physical, science is taken as the model and there is complete rejection of the typology of genes and environment in development.

The implicit assumptions behind IQ tests form the theme of Joanna Ryan's contribution. To psychometrics intelligence is what the tests test; we must therefore see how tests are constructed and what they really do test. Such an examination shows that many social and motivational factors are built into the concept of IQ, indeed they are bound to be because the chief validation of tests is their use in predicting success or failure in schools – and this takes much more than cold-blooded cognition. Looking at the London-line concept of fixed intelligence, the author shows how its apparent validation by the relative consistency in IQ scores is the consequence of the kinds of items included in a test, consistency in the child's social world and a self-fulfilling prophecy effect, the latter stemming from the use of IQ as a means of deciding the kind of education a child shall receive. In summing up, IQ is seen as the rate of change up a scale of difficulty of items that reflect a great deal more than cognition, not at all as the measurement of a fixed quantity of 'intelligence'.

Peter Watson takes up the theme of the social and motivational factors involved in testing and shows how performance is related to the expectation of success or failure. For black children he argues that the expectation of failure is in itself a handicap and produces stress which depresses performance.

The section is rounded off by a broader look at the social and political background of psychometric tradition – John Daniels and Vincent Houghton see the enormous growth of this work coming from two sources – the psychologists' need to gain respectability for their science by mimicking physics, and a desire in the outside world to rationalize selective procedures and to make them appear more democratic. So, for example, in the army, officers should not appear to gain commissions solely because their fathers are gentlemen but because they possess superior 'objectively' measured ability. As psychometrics is used to secure acceptance of one's alloted place in society, the authors argue that it is a political enterprise. In dealing with the casualties of the educational system, they regard testing as the creator of the disease it was intended to cure, as it produces the rationalization that the failures ought to fail. Some of these broader questions are further discussed in the final section of the book.

1 Changing Intelligence
John Radford and Andrew Burton[1]

John Radford is Head of the Department of Psychology, North-East London Polytechnic. He was the Founder and is currently Chairman of the Association for the Teaching of Psychology. He does sporadic research on thinking and conflict resolution, in time snatched between committee meetings.

Andrew Burton began teaching at North-East London Polytechnic in 1968 after graduating in psychology at London University. His research interests include the experimental investigation of brain damage. He is preparing a book on the psychology of thinking with John Radford.

In this paper we take a package tour of some changing views of intelligence, and of attempts to change it. As with other such tours, it is difficult to avoid drawing the reader rapidly past one great work, and ignoring another. Some of the hotels are not even built yet. In the circumstances, the reader is advised to emulate the caution of an American lady who, referring to her guide-book before some Renaissance masterpiece, was heard to remark: 'It says here there are twenty-three angels in this painting. Count them, Sadie!'

The London line

We begin with Sir Francis Galton. Galton arouses our admiration for many things, not least his confident way with a charging lion: '... keep cool and watchful, and your chance of escape is far greater than non-sportsmen would imagine' (*The Art of Travel*, 1855). Galton's travels in Africa showed him the great

1 This paper is based upon parts of a book on the psychology of thinking in preparation by the authors.

diversity of the human race; and when his cousin Charles Darwin put forward his account of *The Origin of Species* (1859) it seemed the next step was to explain the differences between individuals. He tried several tacks, such as fingerprinting and composite photographs, on the way to his eventual discoveries. Galton collected data on the variability of the human species in his Anthropometric Laboratory at South Kensington, in which, for 3d, you could learn exactly your height, weight, breathing power, strength of pull and squeeze, colour sense, and the like. Galton was thus perhaps the only psychologist to have his subjects pay him, rather than the more usual arrangement. However, as is well known, for the study of *Hereditary Genius* (1869) he turned to eminent persons in history and contemporary life.

The received view was that, apart from a few geniuses and a few idiots, all men were equally endowed, and achievements were mainly the result of hard work – 'self-help' in the famous phrase of Samuel Smiles. Darwin himself thought this, as we know from a letter to Galton. Galton started from the now familiar concept of the 'normal' distribution. Quetelet, in 1835, had begun to show that measurements of the human body tend to be normally distributed. Galton argued that this must also be true of psychological traits, since they must, presumably, be based upon physiology. He thus had a distribution for ability. Concentrating on the upper end, he took his original data from the British membership of the *Dictionary of Men of our Time* (1865) and from obituaries of eminent persons in *The Times* for 1868. He put his rank of 'eminent' at one in 4000 of the population. In order to explain the presence or absence of eminence, Galton collected pedigrees of leaders in all sorts of fields – the law, the Church, mathematics, rowing, north-country wrestling. Not very surprisingly, he found a tendency for an eminent person in any one field to be related to other similar persons. From this starting point he went on to argue that superiority was mainly inherited. It now seems naive to belittle the tremendous influence on the individual of social advantage and cultural pressures. Galton's genetics, too, were incorrect in detail. Mendel's experiments (which now form the basis of modern genetic theory) were not rediscovered until Galton was eighty.

The influence of Galton, like that of Freud and Darwin, has been so great that we are now scarcely aware of it. It is really because of his work that we are all now commonly tested at every stage of our lives. The vast enterprise of psychometrics began in that Anthropometric Laboratory. And it could not have proceeded without two statistical concepts of fundamental importance which Galton produced: regression to the mean, to account for the fact that the normal distribution remains stable over time (see chapter 5); and correlation, which besides being the basis on which factor analysis, and thus the whole modern theory of psychological dimensions rests, allowed the notion of partial causation to enter science. Thus one could express the less than complete correspondence between the performance of parents and their children. The question of the inheritance of ability still arouses academic and polemical controversy, and bids fair to continue to do so indefinitely.

It has been one of the issues preoccupying what we have called the London line: Galton, Pearson, Spearman, Burt, Cattell and (in some respects at least) Eysenck. Charles Spearman (1863–1945) was responsible for the famous, and tremendously influential, *two-factor* theory of intelligence. His notion was that any intellectual activity must have two sorts of ability underlying it: one specific to that particular activity, and one common to all such activities. This last was general intelligence or *g*. The idea seems plausible: but where Spearman was too ambitious was in believing that he had analysed the nature of *g*, and that this analysis provided psychology at last 'with its prime requisite, a genuinely scientific foundation' (Spearman, 1925). This foundation consisted of laws or principles, which Spearman had had the good fortune to discover: and 'these principles, (together with commentaries upon them) appear to furnish both the proper framework for all general textbooks and the guiding inspiration for all experimental labours'. After this the actual *neogenetic* principles, as Spearman called them, are something of an anticlimax. They consist essentially in the ability to see relations and correlates; and they are enshrined in every test item which asks 'A is to B as C is to?'. In the work of Burt, *g* developed into 'innate, general, cognitive ability'. Burt believed that he had demonstrated more or less exactly the relative contributions of heredity and environment to test performance. What is quite beyond question is that Burt's pioneering work as a practical psychologist with the London County Council gave intelligence tests an immense prestige. About the same time, on the other side of the Atlantic, tests came into their own with the eminently practical function of sorting out military recruits in the First World War.

Now all this prestige and useful application conspired to give 'intelligence' the reputation of a stable quantity, more or less fixed at birth, which could be reliably and fairly easily measured. Such a view was of course a gross over-simplication of the actual facts and theories. But it seemed in 1961 sufficiently strong for J. McV. Hunt to spend a large part of his book *Intelligence and Experience* pulling it to pieces. While to some extent attacking straw men, he showed with a wealth of experimental evidence – for which you must read his book – that test scores often vary widely during the lives of individuals, and that they could actually, it seemed, be raised or lowered by the addition or deprivation of social and cultural stimulation. He was led to conclude '. . . that it is reasonable to hope to find ways of raising the level of intellectual capacity in a majority of the population'. It is this hope, of course, that has inspired the many cultural enrichment programmes of the last few years. But what is culture? What is enrichment?

Divergence and diversity

Meanwhile, in another part of psychology, the concept of IQ and its testing was in question. It was in 1950 that J. P. Guilford began his now famous presidential address to the American Psychological Association with the words: 'I discuss the subject of creativity with considerable hesitation, for it represents an area in which psychologists generally, whether angels or not, have feared to tread.' (He does not mention Galton.) One might now comment that since then psychologists, whatever they be, have rushed in. A bibliography of 1950 to 1965 by T. A. Razik listed 2088 items; the flood shows little sign of diminishing.

Guilford's work is intended as a comprehensive and ultimately definitive analysis of intellectual functioning. The system, which goes under the name of *the structure of intellect*, begins with a theoretical argument. Any account of intellectual abilities, it is held, must deal with three classes of variables: the activities or operations performed; the material or content on which operations are performed; and the product which is the result of operations. According to Guilford, these variables consist of the following:

Operations: cognition; memory; divergent thinking; convergent thinking; evaluation.

Content: figural (e.g. images); symbolic (letters, syllables, other conventional signs); semantic (meanings, concepts); behavioural (e.g. social perceptions and concepts).

Products: units; classes; relations; systems; transformations.

It is then clear that on this analysis, intellect consists of a large number of separate factors or abilities, viz. $5 \times 4 \times 6 = 120$. There is no general factor. Further, the analysis shows what factors *ought* to exist; their actual existence has to be established empirically. Guilford and his associates present evidence (based on factor-analytic studies of performance over a range of tests) for many but not all of the separate factors. Now this system – which is, despite its name, really an analysis of intellectual *functions* – has several drawbacks. The original analysis rests upon plausible argument rather than proof; it is not too clear that there is corroborative evidence for the multiplicity of factors, outside of Guilford's own statistical results; and the system does not seem to be as usefully predictive as the more general, conventional, 'intelligence'. But it does show up the oversimplification of the older views. In particular, one important point which Guilford made early on, was that conventional tests deal only with one group of abilities. Specifically, they demand 'convergent' thinking: all the tests items are such that only one answer is correct, and all others, even possibly better ones, are wrong. If abilities are to be fully measured, Guilford argued, we must have tests of *divergent* thinking: tests in which credit is given for originality or for a flow of ideas. The notion was not new: apart from Galton, there had been tests of creative ability by Chassell among others. But the time was right: the United States was engaged in a battle of technology with the USSR.

It was essential to capitalize the national resources of inventiveness. Divergence was the thing. As with intelligence itself, the convergent–divergent dichotomy escaped from its maker, and began to lead a life of its own, acquiring in the process evaluative overtones, so that we all began to feel we *ought* to think divergently, and were rather lacking if we could not. Guilford himself never said this, of course, and thinks that creative ability has many very varied aspects. His theory certainly does not include the concept of two sorts of people, creative and non-creative, who think divergently and convergently respectively.

But a number of studies, of which that by Getzels and Jackson is probably the best known, *did* seem to support such a concept. And the demand in industry and society generally for 'creative' people, led to a host of attempts to produce the required product. It is safe to say that it is this area of intellectual activity that has been the focus of by far the greatest number of attempts at modification.

Roughly speaking, the attempts have had either short-term or long-term aims. In the short term, the idea is to stimulate originality in particular situations (which might, of course transfer to other situations). Maltzman, for example, has presented a subject with a list of words to which he is to free associate. The same list is given several times, and the subject tries to give different responses each time. The range of responses, it is reported, increases, and this transfers to other stimulus situations. The more famous methods have involved groups, and of these the best-known is brainstorming. The essential feature of all brainstorming is that of deferred judgement: brainstormers try not to condemn out-of-hand the bizarre but possibly creative thoughts that pop up. But the emphasis on *group* activity is probably misplaced. Bouchard concludes that being in a group actually inhibits creative thinking. These techniques, and the much more complex ones of *synectics*, do not seek or claim to alter a person's basic abilities; but they do suggest that the use of these may be partly dependent upon the demands of the situation.

More ambitious is the attempt to educate for creativity. This has of late become rather popular. There are a number of difficulties. The first is common to all work on creativity, namely that of criteria. What *is* creative? The usual suggestions are that it is something that is novel and/or useful. But these points are less than conclusive. Obviously not all useful things are 'creative', while mere originality is found to excess in the productions of lunatics (and, some would say, of present-day artists). Often enough, in the past, the creative product has at first appeared to be banal, useless, or ridiculous. The trick is to *recognize* the good new idea: but this is precisely what is difficult. Again, the history of both art and science do not give very much encouragement to the notion of making great advances by rule. Rules need to be broken by the creative thinker, as anyone will agree who has read *Hamlet* and *The Spanish Tragedy* (except a confirmed francophile). It is true that advances can be made by sheer hard work in science, which has been described as a means to enable the non-creative to be creative. But that is not the point at issue. The second difficulty is an experimental one: it is a very long and extensive business to discover whether educational programmes result, perhaps many years later, in more creative behaviour. The problems of control are virtually insuperable.

Nevertheless, Parnes and Brunelle, in 1967, reviewing forty such programmes, concluded that students could be reliably taught to improve their sensitivity, fluency, flexibility, originality and elaboration. Does this make sense? Getzels and Jackson argued that conventional schooling tends to inhibit originality, and Liam Hudson agrees. But Hudson studied the particularly restricted environment of public and grammar schools; and he is careful to point out that the effect is more a matter of belief than evidence. Certainly much education, especially in the fourteen to eighteen-year-old range, could be described as highly formal and even rigid. But very successful and creative people manifestly do emerge from it. There is a wide variation even within state systems. One survey reported that over 90 per cent of the questions asked by a sample of junior high school social-studies teachers called only for the reproduction of textbook information. But many excellent, inspiring teachers also exist. Conversely, so-called 'progressive' schools, of which A. S. Neill's Summerhill is probably the best known in this country, are often contrasted with more conventional education.

But we know of no available evidence that they produce more creative adults. Freeman, Butcher and Christie, have reviewed a number of individual schools and teachers allowing greater or less self-expression, and conclude: '. . . that under instructional methods which optimize their abilities creative children enjoy themselves to a greater extent.' This may well be so, and may be all to the good; but it is not *necessarily* linked to future performance.

Now one of the many difficulties with experimental research in education is the extreme, if temporary, malleability of children. New teachers may find this hard to credit, but experienced ones will not doubt that if Sir is pleased with rows of neat sums, that is what he will get, whereas if he prefers poems or puppets, his children will obligingly produce them, and excellent they may be. Hudson has shown that apparent 'convergers' can be just as fluent as divergers when the instructions are unambiguous. '("So *that's* what you want," they seem to be saying. "Why didn't you say so in the first place?")'. And Robert Rosenthal, of course, has reported that teacher's expectations can influence even such ostensibly objective measures as standard I Q tests, without either children or teachers being aware of what was happening. Rosenthal and Jacobson told teachers that some of their children, actually a random sample, 'would show unusual intellectual gains during the year'. Eight months later the test scores of at any rate the first two grades were, apparently, significantly higher. Though this study has been widely criticized on statistical grounds, its general conclusion fits the views of many teachers and educationists.

Accordingly, it is not too surprising when success is reported in altering much more flexible behaviour. Richard Crutchfield believes that in our rapidly changing society, we can no longer hope to equip children with specific facts, or even skills, that will serve throughout their lives.

What education must ... seek to do is bring about the optimal development of the whole individual. He must be equipped with *generalized* intellectual and other skills, skills which will enable him to cope effectively with whatever the state of the world is as he will later encounter it. . . . Central among these generalized skills is the capacity for creative thinking.

To train this capacity Crutchfield has prepared a series of sixteen booklets, in which two children, Jim and Lila, aided by their kindly Uncle John, work their way through a series of problems, such as strange happenings in a haunted mansion. Jim and Lila are led to generate hypotheses, to check these against facts, then reformulate the problem, and so on. The reader, for his part, tries to keep one step ahead of the characters. 'Mainly, the aim is to build up the child's successful experience in coping with thought problems.' At the same time he is trained in techniques likely to bring this about, and in favourable attitudes to creative, flexible thinking, Crutchfield's 1965 report dealt with 267 ten- and eleven-year-old experimental subjects, and 214 controls. Pre- and post-test procedures included measures of intelligence, personality, cognitive style, creative ability, and self-evaluation as a creative thinker. Children at all levels of intelligence surpassed the controls after training and produced more and better ideas, and this effect persisted six months later. But there was little change in attitudes, and least of all in children's self-confidence as creative thinkers.

Even without Rosenthal's experiments, it would seem plausible, if you wish to change children in some way, to start with their teachers. Several investigators have done so. Torrance pertinently asks: 'How . . . can one explain why certain teachers produced so many students who made outstanding discoveries?' He suggested a number of 'workshops' to aid teachers in developing relevant skills. Barron used some of Gordon's synectics techniques with teachers in Californian state schools who were encouraged to think in a 'new and imaginative way' about problems such as marijuana and prostitution. They got as far as considering legalization, and showed 'significant gains' on measures of creative thinking.

There are many other examples of similar kind. We confess to a measure of scepticism. By all means let teachers, and children too, explore new possibilities, but just what is it all *for*? Is there any evidence at all of permanent effects, or of an increase in productivity in later life, which is presumably what is intended? Education – that is, universal school education – is culturally and historically a very unusual phenomenon, as Bruner points out. All the great achievements of the past occurred without it.

Will anyone claim that the Greece of Pericles, or the England of Shakespeare, were less creative than our own time? One might say there is a vast confusion between value judgements and the technical means to an end. It is an error to which functionalist psychology has been prone; even when the means do not necessarily work. Happy children are one thing. 'Creative' children may be quite another; and creative adults a still more distinct third. How happy was the childhood of Charles Dickens? Even at a level more amenable to investigation, Astin failed to find any evidence that a student's achievements were facilitated by the intellectual level of his classmates, the level of academic competitiveness, or the financial level of his institution.

A living organism

At the time when the testing movement was getting into full swing, a Swiss biologist with leanings towards philosophy began to be interested, not so much in how many children gave the right answers, as why they made the particular mistakes they did. Jean Piaget, who was already a productive zoologist, turned his attention to psychological problems, and he spent some time applying Cyril Burt's tests to French children. Burt's tests embodied Spearman's principles of seeing relations and correlates. But how did children come to be able to do this? In an autobiographical article, Piaget says: 'While I wanted to devote myself to biology, I had an equal interest in the problems of objective knowledge and in epistemology.' (Epistemology: the theory or science of the method or grounds of knowledge, *Oxford English Dictionary*). 'My decision to study the development of the cognitive functions in the child was related to my desire to satisfy the two interests in one activity. By considering development as a kind of mental embryogenesis, one could construct a biological theory of knowledge....'

Such an attempt was quite out of tune with the psychology of the 1920s, when Piaget began to publish. His work was highly theoretical, not practical, like testing; it was biological, not clearly based on mechanical stimuli and responses; it talked about mental processes when Watson had claimed these to be scientifically disreputable, if not actually non-existent; his experimental work, such as it was, lacked all the proper controls.

Due to such beliefs Piaget spent forty years in the psychological wilderness. Perhaps a more fundamental reason was that he has always been interested in *structure*, rather than in *function*, which had dominated American psychology since William James. Just now, of course, Piaget has been brought back from the wilderness and is received everywhere with honour. This is indeed his due: but it has not prevented the continuation of mistakes about his work. (Structuralism as a whole, of course, is enjoying a vogue, in mathematics, linguistics and anthropology, besides psychology.)

We can ignore the mistakes about lack of experimental sophistication, small numbers of subjects, and so on. There are now, besides, many good clear accounts of the main stages of intellectual development which students used to find so puzzling. Let us rather, if we can, try to see what it is that Piaget has done that is so novel, and why he has enabled us to see intellectual activity in a completely new way. It is perhaps not too much to claim that he has produced not just a new theory, but a new area of science. Until very recently, the question of how knowledge comes about has been a matter for argument rather than a question of fact. One of the main issues was whether knowledge is inborn, or whether all we know is acquired through experience. The former line was taken by the philosophical predecessors of the Gestalt school, among others; and the latter by the British empiricists whose heirs were the strictly environmentalist behaviourists. The argument over the inheritance of intelligence was part of this issue, which, it appeared, was an either/or matter. In practice, of course, since it is pretty obvious that intelligence could not be *entirely* due only to heredity or only to environment, the answer usually given has been some percentage of each. By bringing the basically philosophical argument within the scope of empirical investigation, Piaget has shown not so much that that is the wrong answer; it is an answer to the wrong question.

First of all, Piaget's approach is a biological one. Despite the fact that man is obviously a living organism, for many years the physical sciences prevailed as models for psychology (e.g. the stimulus–response 'chemistry'; Gestalt 'physics'; ethological hydraulics; Freudian dynamics). All such views essentially see man as a very complex machine, responding to stimuli in a mechanical way. For Piaget, it is only meaningful to speak of behaviour as a two-way interaction between the organism and its environment. Consider man, his brain and his mind, as a tree: its growth can only be understood as a process of interaction between the tree and the nutrient or harmful aspects of the environment. One thing to follow from this is that the organism is essentially organized: not just a set of parts making up a whole, but an entity all of whose aspects are essential to the others. Another point is that in biological development, there is as it were no zero point, no point before which the organized interaction with the environment is lacking. When does a tree begin its life? – with the seedling, or the seeds, or the parent flower? Rather, there is always a structure, evolving by interaction with the environment into a more complex structure. This is Piaget's answer to the nature–nurture issue. He develops it through the immensely complex and detailed theory of stages. As to what intelligence *is*, Piaget rejects any account of it as a set of capacities – such as Burt's, or Spearman's, or Guilford's. 'I would define intelligence,' he says, 'as a form of equilibration, or forms of equilibration, towards which all cognitive functions lead.' And then: 'I define equilibration principally as a compensation for an external disturbance.' Furth, puts it thus: '... intelligence is the totality of behavioural coordinations that characterize behaviour at a certain stage ... intelligence is the behavioural analogue of a biological organ which regulates the organism's behavioural exchange with the environment.'

The organism interacts with the environment, in Piaget's view, in certain fundamental ways, termed functional invariants. The first we have already stressed: it is organization. The second is adaptation, which has two complementary aspects, assimilation and accommodation. Assimilation occurs whenever the organism utilizes something from the environment and incorporates it – food, stimulation of any appropriate kind. Accommodation is the process of changing the organism's internal structures to suit and make use of the input from the environment. Now we might expect that Piaget would be particularly interested in the environment, since he lays so much stress on it. In fact, he has tended to treat it merely as a necessary condition.

We might expect his theory to tell us exactly what in the environment favours intellectual development, as a biologist might advise a horticulturist. But this, it seems, is not as yet possible. We might also expect attempts to be made to hasten development; and this, in fact, has happened (e.g. by Sigel and Hooper, in 1968). But Elkind seems clear that such attempts are mistaken. We think they are premature rather than wrong in principle. If Piaget is right, it must ultimately be possible to specify environmental variables that will favour or inhibit intellectual growth. But that will be a far cry from the sort of short-term training programme to speed up the transition from one stage to another that has so far been attempted. The lesson of Piaget's work, rather, is that intellectual functioning is not something like the performance of a motor car, a combination of a piece of engineering with oiling, fuelling and servicing, but a continuous process of individual growth. And 'intelligence' is not a set of discrete capacities.

Culture

What, then, can be said of the role of the cultural environment in which the individual finds himself – the family, education, society? It has been perhaps too easy to assume that because IQ scores are somewhat variable, not fixed immutably, and because we can encourage children to think 'creatively', it will therefore be an easy task to raise the general level of intelligence. It has seemed clear that we should get rid of intellectual handicaps as we have got rid of rickets. But it is not so simple. According to L. J. Cronbach:

> The phrase 'improve the environment', born of the enthusiasm of the Social Darwinists, has misled environmentalists for two generations. Environments cannot be arrayed from good to bad, rich to poor. The highly stimulating environment that most of us think of as 'rich' promotes optimal growth for some persons and may not be suitable for others.

Nevertheless some general lines seem fairly clear. We know, from Hunt and many others, that on the whole an urban environment and a more sophisticated milieu, tend to be accompanied by higher IQ scores. Extensive studies have clearly shown that institutionalized children, lacking the variety and individual attention of home life, are at a disadvantage, especially in language development. P. E. Vernon studied West Indian children in a variety of home and school situations. Here, he says, '. . . it is clear that the most important single factor in children's performance on *g* and verbal tests is the cultural level of the home, parental education and encouragement, reading facilities and probably the speech background.' Vernon's name should really be added to what we have called the London line. In 1969 he presented a comprehensive review of his own and other work on the effects of culture. Taking his cue from Hebb he distinguishes three aspects of intelligence: *A* which is the genetic potential; *B* is 'the effective all-round cognitive abilities to comprehend, to grasp relations and reason, which develop through the interaction between the genetic potential and stimulation provided by the environment'; *C* is test scores – really a sample of *B*, if the tests are good ones. Now what promotes *B*? Vernon is clear: '. . . it seems reasonable to regard the Puritan ethic of the western middle class as producing the greatest development of intelligence.' He is clear too about what to do for underdeveloped countries: '. . . the greatest promise of quick advance lies in the field of language-teaching' – i.e. a sophisticated language suitable for abstract and technological thinking, such as English. We should not from this conclude more than the fact that English happens to be the language in which more sophisticated things are said; it is not necessarily more complex or 'superior'.

There are some clues, exciting but perhaps misleading, as to more specific effects. For example Carlsmith, in 1964, tested Harvard students whose fathers had or had not been absent overseas during their childhood, due to the Second World War. Since this was an absolutely objective fact, and one over which the fathers had virtually no control, it was an exceptionally elegant natural experiment. There was a clear tendency for early and long separation from the father to be related to greater verbal than mathematical ability. Somehow, it seems, fathers foster mathematical aptitude. Vernon, finding that Ugandan children did poorly on a speeded form-board test (fitting triangular pieces together to make different shapes), speculates as to the effects of spending the first year or two of life bound to the mother's back. Thus the child receives little manipulative or kinaesthetic experience, while vision is dominated by one rounded object, the mother's head.

The furthest explorations along this line are probably those of Bruner and his colleagues. They take a view of intellectual growth inspired by but not identical with that of Piaget. They stress the differing demands of technologically advanced and backward societies, necessitating different *modes* of thinking. Maccoby and Modiano (1966), for example, compared Mexican children from rural and urban environments. The first group were poorer at Piagetian tasks involving, essentially, the ability to manipulate the environment in one's head, to think abstractly. Such thought is not required of peasant farmers, whose life depends rather on concrete, perceptible facts: 'his crops, the weather, and particular people around him'.

Cross-cultural studies are fraught with difficulties. Price-Williams (1961), for example, found with the Tiv people of Nigeria that their performance on Piaget-type tasks was approximately equivalent to that of Swiss children, provided that familiar materials, such as native acorns, were used. Far too often, in the past, a grossly oversimplified approach has been used. A leading example here would seem to be the admittedly ingenious attempt of D. C. McClelland to account for intellectual and economic advances in terms of the level of 'achievement motivation' in a society. And even if we can, eventually, isolate exactly the important cultural variables, there will still be the problem of the interaction of these with *individual* genetic endowments.

We can agree with Holtzman that psychology should lose its ethnocentric character. But it is ironical that we become aware of this at a time when many cultural variations seem about to disappear for ever. And this too should give us pause in our attempts to change intelligence. There are human societies, in South America, Australia and elsewhere, well adapted to their environments, who will shortly suffer the fate of the Tasmanians. Thus we lose a precious reservoir of genetic variability. No less valuable is cultural heterogeneity. From such diversity, it would seem, sprang most of the human accomplishments of which we are proud. If we can already, in broad terms, change intelligence, by what criteria are we right to do so? As knowledge becomes more exact, the question becomes more pressing.

And so, perhaps, our tour comes full circle. For Galton's unfulfilled ambition was to bring about, by the science of eugenics, the betterment of the human race. Genetic engineering is now within our grasp; many people would find this morally objectionable. But what of cultural manipulation? Here too Galton was a model of Victorial certainty:

The best form of civilization in respect to the improvement of the race, would be one in which society was not costly; where incomes were chiefly derived from professional sources, and not much through inheritance; where every lad had a chance of showing his abilities and, if highly gifted, was enabled to achieve a first-class education and entrance into professional life, by the liberal help of the exhibitions and scholarships which he had gained in his early youth; where marriage was held in as high honour as in ancient Jewish times; where the pride of race was encouraged (of course I do not refer to the nonsensical sentiment of the present day, that goes under that name); where the weak would find a welcome and a refuge in celibate monasteries or sisterhoods, and lastly, where the better sort of emigrants and refugees from other lands were invited and welcomed, and their descendants naturalized.

In the phrase of the well-known examination question: Do you agree?

Further reading

J. S. Bruner, R. R. Olver and P. M. Greenfield *et al.*, *Studies in Cognitive Growth*, Wiley, 1966.

J. Freeman, H. J. Butcher and T. Christie, *Creativity: A Selective Review of Research*, Society for Research in Higher Education, 1968.

H. G. Furth, *Piaget and Knowledge*, Prentice-Hall, 1969.

F. Galton, *Hereditary Genius*, Macmillan, 1869.

J. W. Getzels and P. W. Jackson, *Creativity and Intelligence: Explorations with Gifted Students*, Wiley, 1962.

L. Hudson, *Contrary Imaginations*, Methuen, 1966; Penguin, 1968.

L. Hudson, *Frames of Mind: Ability, Perception and Self-Perception in the Arts and Sciences*, Methuen, 1968; Penguin, 1970.

J. McV. Hunt, *Intelligence and Experience*, Ronald Press, 1961.

A. F. Osborn, *Applied Imagination*, Scribner, 1957.

J. L. Phillips, *The Origins of the Intellect: Piaget's Theory*, Freeman, 1968.

P. E. Vernon, *Intelligence and Cultural Environment*, Methuen, 1969.

S. Wiseman, *Intelligence and Ability*, Penguin, 1967.

A full list of references may be had on application to the authors.

2 IQ - The Illusion of Objectivity

Joanna Ryan

Joanna Ryan is a Fellow and Tutor of King's College, Cambridge. Her research work, carried out in the Unit for Research on the Medical Applications of Psychology, concerns language acquisition and mental subnormality.

Psychologists have developed a large number of different tests to assess cognitive abilities of all kinds, in particular, intelligence. In most cases the aim has been to produce some form of quantitative measurement that expresses an individual's standing relative to others, on a one-dimensional scale. Attempts to construct quantitative and standardized tests of intelligence have been going on for the last seventy years, mostly in western countries and there is still a large number of psychologists engaged in refining old tests or developing new ones. Currently, the government's Department of Education and Science is giving substantial financial support for the development of a revised and thoroughly standardized test of British intelligence. The construction of intelligence tests, as well as other cognitive and personality tests, forms an important subdivision of western academic psychology, this subdivision being known as psychometrics. In other countries test construction is much less important. This is not due to any lack of interest in psychology or education, but rather to a radically different intellectual and social framework.

The purpose of this chapter is to give a critical account of what is involved in this enterprise of trying to measure intelligence, in order to explain what IQ scores mean. It is important to clarify the meaning of IQ scores before considering, as in later chapters, the differences that are found between groups of people. IQ scores are very complex in the sense that many assumptions and operations are involved in their calculation, and it is only when the complexities are fully understood that it is legitimate to try to explain the origin of differences in scores between people. Here we shall consider what it is that intelligence tests purport to measure; what assumptions are made about intelligence in constructing intelligence tests, and how IQ scores can be interpreted. This chapter will concentrate on the aims and assumptions common to the few most widely used tests, and will not include any discussion of the details of the many different and more specialized tests that now exist. Such discussion can be found in a book by Anastasi.

Intelligence

In what follows 'intelligence' will be used to refer, in a general way, to the kind of behaviour the tests were intended to assess. 'IQ' will be used to refer to the scores obtained from such tests. The main focus of this chapter is the question of what the tests do in fact measure, and thus what IQ scores mean. However, it is necessary to discuss briefly what 'intelligence' is commonly taken to be, and also its various definitions, as provided by psychologists.

We all have a rough but unanalysed idea of the difference between people who are clearly mentally deficient, and those who are clearly not. Many of the differences between such extreme groups can be called differences in intelligence. People who are extremely slow in developing, who cannot perform adequately in school, who need 'special' education, who cannot earn their own living, or only under sheltered conditions, who cannot live independently without some form of care – such people are often said to be less intelligent than those who can do all these things. Mental deficiency is very complex in the sense that there is a large variety of causes, as well as many cases where the causes are unknown. Further, some people may not be able to do the things listed above, and thus be labelled as mentally deficient, because of primarily emotional rather than cognitive disabilities. However, whilst there is considerable argument about the precise criteria for defining mental deficiency, and especially about where to draw the line between 'defective' and 'normal', we do still have a reliable idea of how people who are clearly mentally deficient differ from those who are said to be of normal intelligence.

Psychologists have tried to provide us with a more precise and analysed idea of intelligence than this rather vague one. They have also hoped to draw finer distinctions in intelligence, so that people can be graded into many categories, not just the extreme ones. One influential notion has been that of 'adaptability' to the environment. The ability to adapt, and thus to survive, is clearly something which most mental defectives lack. They cannot adapt their behaviour to the prevailing demands of society, and they would not survive unless cared for. 'Adaptation' however has biological and social connotations which most psychologists have wished to ignore in their definitions of intelligence. In general, psychologists have tried to isolate and study only the cognitive components of adaptability. It is one argument of this chapter that this is in principle impossible, and that all attempts to do so are, in fact, implicitly influenced by social criteria. Some of the better known and influential definitions of intelligence, or intelligent activity, are as follows: 'to judge well, to comprehend well, to reason well, these are the essential activities of intelligence' (Binet and Simon); the ability 'to carry on abstract thinking' (Terman); 'the aggregate or global capacity of the individual to act properly, to think rationally, and to deal effectively with his environment' (Wechsler); 'intelligent activity consists in grasping the essentials in a situation and responding appropriately to them' (Heim); and finally 'innate, general, cognitive ability' (Burt). It can be seen that these definitions emphasize abstract reasoning ability, and this is a common theme in many tests. Other psychologists have felt that it is impossible to give a precise definition of intelligence, and have instead maintained that intelligence 'is what the tests test'. This, of course, begs the whole question of what *do* the tests test, and it is this problem we will now consider.

Background

The pressures and traditions that influenced the development of intelligence tests were both practical and theoretical. The practical pressures were, and are, mainly educational. The first intelligence tests were developed by Binet for the Paris education authorities to separate the 'dull' children who would not benefit from the education then offered and who needed special education. Much of the form and content of Binet's original tests is retained in the modern and widely used Stanford–Binet test (Terman and Merrill, 1960) which will be discussed in more detail below. Separation into special schools and training centres for the educationally and severely subnormal is still an important use of intelligence tests today. They are also, in various forms, and sometimes with different names, used in selective examinations at eleven. They were an important ingredient in the old 11+, and are now, under the guise of 'scholastic aptitude' tests, used to determine streaming within some comprehensive schools.

Apart from such demands from the educational system for efficient selection devices, the construction of intelligence tests has been a considerable theoretical problem within academic psychology. Here there is a long tradition of interest in the description and measurement of complex human characteristics, and of individual differences in these. Particularly in the fields of intelligence and personality it might be felt that the technique of test construction has become an end in itself. Internal problems of measurement, such as the consistency of test items with the total score, and the transformations performed on the same scores to give desirable statistical properties to the final scores, are a major preoccupation of many psychologists engaged in test construction. This preoccupation with the techniques of measurement has meant that relatively little attention has been paid to the substantial content of tests, and to their validity as tests of intelligence. 'Validation' is part of psychometric jargon; it involves proving that a particular test does in fact measure what it was designed to assess, and not some other aspect of behaviour. This is usually done by correlating scores on the test with some other criterion of the ability or behaviour allegedly assessed by the test. The second criterion has to be quite external to and independent of the items in the test, otherwise the procedure is circular. Considerable problems arise in deciding what external criteria to use in the validation of intelligence tests. There is some discussion amongst psychologists about this, but considering how crucial this question is to the interpretation of the IQ scores, this discussion is inadequate in all respects, reflecting the confusion about defining intelligence and also the somewhat *ad hoc* and atheoretical criteria used in the selection of items for inclusion in tests (see below). Some test constructors make no attempt at validation, instead relying on a statistical analysis of the test scores. Some correlate their tests with other well established ones, and others 'validate' their tests against school or academic performance or even subsequent social status. Educational success is, in fact, the most common external criterion used. As Butcher (1968) somewhat ironically remarks '... in practice, test constructors, *with the confidence of practitioners in a well charted area*, are generally content with specific criteria, such as correlation with a well-established test or adequate prediction of scholastic attainment' (my italics). This attitude to validation is the source of much mystification about what IQ scores mean, mystification which is unfortunately increased by the internal technicalities of quantification that give intelligence tests a largely spurious appearance of scientific respectability.

What are the tests intended to measure?

Educability or educational success

Since much of the motivation for the construction of intelligence tests came from the demands of the educational system, it might well be thought that the tests measure something to do with educability – that is, the ability to benefit from the education offered. Such an interpretation of intelligence tests is reinforced by the fact that the main external criterion used to validate tests, is, as mentioned above, educational achievement. Validation against school performance was started by Binet who ascertained that scores on his tests did in fact differentiate in the expected direction between children who were thought from classroom observations to be 'bright' and 'dull'. Later psychologists have used more formalized expressions of educational performance, such as examination results, to validate their tests. Furthermore, there are many studies showing that well established tests do predict school achievement with considerable accuracy.

One would thus have considerable justification for supposing that intelligence tests measure educability, or something closely associated with this. However, as the next section will show, psychologists set out to measure something very different from simply the likelihood of educational success. It should be pointed out that in as much as the tests do measure educability, then they are measuring something which is extensively influenced by social and motivational factors of many kinds. A child's ability to profit from school education will depend not only on his cognitive abilities, but also on the kind of education offered, the teachers and their values and abilities, the child's home background, and many other primarily cultural factors. This has been documented extensively in, for example, Douglas (1964), the *Plowden Report*, and Pidgeon (1970). Complex social factors of this kind are certainly not what the psychometrists were after in the construction of intelligence tests. However, in attempting to validate IQ tests by correlating test scores with educational and social success, many important social influences are thereby implicitly introduced. IQ tests do not, and could not, assess only the cognitive, as opposed to social or motivational, determinants of school success. As is argued in greater detail below, cognitive abilities cannot be studied in isolation from their social and motivational determinants, and it would remove much of the mystification surrounding the interpretation of IQ scores if this were explicitly recognized.

Innate potential

One influential but confused theme behind much of the work on intelligence tests is that of genetically determined potential ability. This involves the notion of ability that is a characteristic of an individual prior to any interaction with the environment, and thus independent of any social or specific educational influences. Potential ability is contrasted with actual or current achievement, usually favourably. It is, in theory, ability as determined only by the genotype (see chapter 6) and represents what an individual would in principle be capable of in an ideal, but unspecified, environment. Many tests were designed with this vague idea of intelligence as 'innate potential' behind them, though in differing and often qualified forms. One particularly influential view of intelligence as innate potential is that of

Burt: 'we may safely assert that the innate amount of potential ability with which a child is endowed at birth sets an upper limit to what he can possibly achieve at school or in after-life'.

There are several reasons for supposing that it is in principle impossible to measure 'innate potential', and also that the notion itself does not make sense. The main reason stems from the fact that in the process of measurement some aspect of the individual's current behaviour has to be used – that is, some of the skills that develop during a lifetime. This is because potential is necessarily expressed in actual behaviour; and there is nothing extra 'behind' the behaviour corresponding to potential that could be observed independently of the behaviour itself. Whilst acknowledging that in principle potential can only be measured via its expression in actual behaviour, many test constructors have still tried to devise tests that reflect as nearly as possible an individual's potential ability rather than his current achievements. They try to do this by minimizing the dependence of their tests on particular educational or social experiences, using skills and knowledge that are assumed to be common to all. Intelligence tests are in content very different from school attainment tests, and to this extent they do discount some specific aspects of different educational experiences. However, they do not, and cannot, minimize any more general effects of education and upbringing. The cumulative effects of different social histories are extremely complex and pervasive, and an individual's behaviour will always reflect this (see chapters 8 and 9). Any behaviour, such as performance in an IQ test, that is used as a measurement of an individual's potential, or as an alleged approximation to this, will therefore be influenced by many complex and cumulative effects of this kind. Thus the notion of potential ability both as something abstracted from all interactions with the environment and at the same time as something measurable in a person's behaviour simply does not make sense.

None of this is to deny that there are genetic as well as environmental determinants of cognitive ability. Nor is it to deny there are limits to the extent to which environmental variation can influence an individual's performance and that these limits may in part be determined by a variety of genetic and constitutional factors. What is asserted is that it is impossible to separate out and measure with a behavioural test only the non-environmental determinants of ability, since these interact with the environment in such a was as to ensure that any test of ability will inevitably involve both aspects. Even where tests make minimal use of acquired educational skills, performance on the test will always be influenced in a profound and complex way by the experiences of a lifetime.

The idea of potential is asymmetrical in the sense that potential is seen as setting upper but not lower limits on ability. To some extent this asymmetry corresponds to our present knowledge concerning environmental manipulation. Whereas we do not at present know how to change the abilities of, for example, mongol subnormals to reach the 'normal' range, we do on the other hand know of a large variety of ways in which ability can be depressed. Further, there is probably no limit, including death, to the extent to which someone's ability can be lowered. However, this apparent asymmetry may simply reflect our partial knowledge of means of manipulation rather than being an unalterable fact of human nature, as implied by Burt's definition of potential. Methods for interference with the genetic material are no longer pure fantasy. The success of dietary control in treating mental subnormality due to gene abnormalities, as in phenylketonuria, further revises our idea of genetics as fate.

Linked with the notion of innate potential is the idea that the intelligence of an individual is 'fixed', and does not change throughout life. All intelligence tests are constructed on the assumption that intelligence is a stable characteristic of the individual throughout life. This assumption is of course implicit in the definition of potential ability as something unaffected by interaction with the environment. It follows from this assumption that an individual's IQ score should be constant throughout life, and the tests are designed by suitable selection of items to give as much constancy as possible.

Psychologists have devoted a lot of attention to the question of whether IQ scores do in fact remain constant throughout life, and thus whether early IQ scores are good predictors of later ones. In general, individual IQ scores appear to change less with increasing chronological age. Thus below the ages of four or five they are very unreliable as predictors of later IQ and even after this there are many cases in which large changes have been recorded. There are very many different factors which might determine the stability or otherwise of IQ scores, and they cannot all be discussed here. However, three points can be made. One is that even if it were shown that IQ scores were highly stable, this would not mean that they therefore measured innate potential or indeed innate anything. Any one of the determinants of IQ scores could well produce the apparent stability, including the most obviously environmental determinants. There are many relevant environmental events that have early, cumulative and persistent effects on cognitive abilities, and which could therefore produce apparent stability in IQ – for example, parental encouragement and interest. Secondly, IQ testing may well be a self-fulfilling prophecy in many cases. For example, IQ scores obtained at one age often determine how an individual is subsequently treated, and, in particular, what kind of education he receives. The kind of educational treatment an individual receives as a consequence of IQ testing will in turn contribute to his future IQ, and it is notorious that those of low and high IQ do not get equally good education. Differential treatment of this kind will thus tend to produce stable IQ scores. Thirdly, most IQ tests are constructed on the assumption that intelligence is an unchanging characteristic of an individual, and much effort has been put into finding suitable test items that will yield constant IQ scores. The constancy of IQ scores may thus be more a fact about tests than about people.

In this section we have discussed the assumption that individuals possess, as an innate and stable characteristic, potential ability that can be measured with intelligence tests, albeit approximately. This assumption is implicit in nearly all work on intelligence testing, although it is not always made explicit. We have argued so far that this assumption, that potential exists and can be measured, does not make sense. What is worse is that this assumption is insufficiently acknowledged in some of the uses to which intelligence tests are put, leading to some paradoxical and dangerous results. One example of this is seen in their use to assess the effectiveness of educational intervention programmes, such as Headstart (see Conclusions: Intelligence and Society). Here the IQs of the children before and after the intervention are measured, and any increase in IQ is taken as an indication of the success of the programme. It seems quite wrong to attach so much importance to a change – or to an absence of change – in IQ score, when the test is designed both on the assumption that what is measured is a fixed and unchanging characteristic, and with the aim of producing constant IQ scores. Such a use of intelligence tests suggests insufficient appreciation of the assumptions implicit in their construction.

General intelligence

There has been much debate amongst psychologists about the 'structure' of intelligence. This debate has taken the form of asking whether behaviour in an intelligence test can be accounted for in terms of many unrelated and specific abilities or by one general cognitive ability. Examples of specific abilities would be verbal, spatial, perceptual, motor and numerical abilities. The question at issue is whether there is any correlation between one of the abilities and any of the others. Do people who do well at verbal tests also tend to do well at all others, and do people doing badly on verbal tests also do badly on all others? Those who believe in the general nature of intelligence believe that these specific abilities are highly correlated, whilst those who believe in its particulate nature, hold that these abilities will tend not to be intercorrelated.

There is no definite answer to this question, because the answer obtained depends on the structure of the tests used, and the population sampled. However, as far as the most widely used tests of intelligence are concerned, namely the Stanford-Binet and the Wechsler scales, the answer seems to be a compromise. These tests are heterogenous, in the sense that they require many different specific abilities. It is found that performance on different kinds of items is quite highly correlated, but not totally. Such tests therefore assess ability of a fairly general nature, in that this ability determines the level of performance on many different items. However, specific abilities are involved to a small extent.

So far we have discussed three ideas – educability, innate potential, and general intelligence – in relation to the question of what intelligence tests purport to measure. We have concluded so far that they do assess something closely related to the likelihood of educational success, but that this cannot be 'potential' of any kind since this idea does not make sense. Further IQ tests involve cognitive abilities of a very varied kind, and the development of these abilities is influenced in a complex way by all possible environmental events. We will now consider in more detail the content of the tests, and their construction and standardization. This will give a more specific and concrete picture of what they contain, and thus what IQ scores mean.

What kind of measuring instruments are intelligence tests?

Before discussing the substantial content of tests, it is necessary to describe the kind of measuring scale they provide. Despite the fact that scores on all such tests are expressed numerically, the tests do not measure absolute quantities, nor do they have a zero point. In this, as in many other respects, they are very different from familiar measuring instruments such as rulers and thermometers. In intelligence testing, individuals are ordered, or ranked, according to how well they do on the test. Scores on such tests represent the position of individuals relative to each other, or to a comparison group. These scores, however, do not convey anything about the size of the differences between individuals in different positions. From any two scores all we know is that one individual has done worse or better than another, and we can order large groups of individuals in this way. We cannot treat the numerical test scores arithmetically in the same way as we can the figures obtained from measuring the height of people. Thus we cannot say that someone with an IQ of 100 has twice the IQ of someone with an IQ of 50, whereas we can say that someone of 6 ft is twice as tall as someone of 3 ft. Further, equal numerical differences in IQ (e.g. between IQs of 50 and 100, and of 100 and 150) do not mean that the differences between individuals with IQs of 50 and 100 resemble the differences between individuals with IQs of 100 and 150 in any respect at all. We cannot use IQ scores arithmetically, because the IQ is in the form of what is known as an ordinal, and not an interval, scale. This point has been laboured somewhat because the numerical precision with which IQs are expressed can give the false impression that an IQ scale measures intelligence in the same way as a ruler measures length. This false impression gains credence from the common practice of calculating average IQ scores for groups of people from individual scores, and of using the averages in experiments. This practice is quite unjustified, for the reasons advanced above.

The selection of test items

We shall now consider what criteria are used in the selection of items for inclusion in intelligence tests. As mentioned earlier, the items in most tests are very varied and cover a fairly wide range of abilities. For example, the Stanford–Binet test at a six-year-old level contains the following kinds of items: vocabulary, comparison of pictures for similarities and differences, completion of pictures, number concepts, verbal analogies and maze tracing. Many items, especially at older age-levels, presuppose skills such as reading or writing, or ask for specific information. Others assume certain values or preferences on the part of the child, and many contain obviously middle-class social and cultural biases (e.g. 'why do we have books?'). The important question of what kind of interest and motivation the use of such items and the administration of tests in general presuppose, is discussed by Peter Watson in the next chapter.

The basic requirement of an intelligence test covering a wide chronological age range is that there should be a scale of difficulty on which individuals are graded. In the Stanford–Binet test this is achieved by including items that themselves vary in difficulty, and assigning these items to different chronological age ranges. The items start at a level roughly suitable for two-year-old children and increase in difficulty up to a level suitable for sixteen year olds. The items are grouped into discrete levels, six items per level, the levels being arranged at either six-month or one-year intervals. This provides a scale of increasing difficulty, on which a child's performance may be assessed, regardless of his age. In order to construct this scale of difficulty the test constructors have first to determine which items are in fact more difficult than others. They do this by choosing items which will differentiate approximately between most children of different chronological ages. Thus an item selected for inclusion at a particular age level, for example, five years, might be passed by 50 per cent of all children tested at five, and by 90 per cent of all six year olds. These figures are purely hypothetical and the percentage criterion that is used in practice in the construction and revision of the Stanford–Binet test varies from item to item and level to level. The scale of difficulty is thus constructed in terms of the proportion of children in the construction sample who pass the items at each age.

The Wechsler test for children (WISC) also contains an age-related scale of difficulty against which any one child's performance is assessed. Unlike the Stanford–Binet test the items are the same for all ages. The scale of difficulty is constructed from how well children of different ages do on the same item. The items are thus graded internally for difficulty, and children are compared in terms of typical scores for any one age level.

The basic principle in both the Stanford–Binet and the WISC is however the same: the test items, or the scores on these, are chosen to reflect the abilities of children as these change with age. The primary operation involved in setting up the tests is to differentiate children in one age group from those in other age groups. The main criterion for selecting test items is that they should show differences of an appropriate statistical kind between age groups. In the various revisions of the Stanford–Binet test (by Terman and Merrill) this basic criterion is refined and extended. Greater stringency is exercized in the percentage criterion and its spread over several age groups – very sudden changes with age are avoided. Another statistical criterion used is the contribution that any one item makes to the total score; that is, the correlation of scores on that item with the scores on all the other items combined. The aim here is to ensure that all items correlate highly and to the same extent with the final score. This is known as the consistency of the test. A further criterion employed is the reliability of the items; the extent to which the same results are obtained over a period of time on re-testing.

A somewhat different criterion is the absence of sex differences. Items showing large or consistent sex differences are excluded. This fact vitiates all attempts to show sex differences in ability by use of intelligence tests, as is sometimes done.

What have these selection criteria to do with intelligence?

The selection of items is not influenced by any theoretical view of how children develop with age, nor by any considerations as to the developmental significance of the particular items. There is no discussion in the manuals for these tests of why a scale of changes in ability with chronological age should be chosen to measure intelligence. Nor is there any discussion of the relationship between the particular items and the assumed nature of intelligence. Because no attention is paid to the developmental significance of the test items, there is no guarantee that some of the items do not simply reflect rather trivial aspects of behaviour. It cannot be assumed that any change in ability with chronological age is an important one as regards intelligence. Some measured changes may well be relatively trivial consequences of other changes, for example, of growing larger and stronger.

Thus the criteria that determine the selection of test items are purely formal, and relate to the requirement of the test as a measuring instrument, and the desirable statistical properties of this. The selection of test items is not related to any views of how children develop, but is determined by quantitative criteria alone. Finally, no attempt is made to explain how a developmental scale of difficulty is related to what the tests were designed to measure, viz. intelligence. This point is elaborated in the next section.

The meaning of IQ scores

An IQ score is essentially an expression of how far up the scale of difficulty an individual has got with respect to his chronological age group. For example, a child of five years who has reached the six-year-old level will have a high, above average, IQ since he has got further than most of his peers. A child of ten who has reached the nine-year-old level will have a low, below average, IQ score, since he has not gone as far as most of his age group. Thus an IQ score expresses an individual's rate of change on the scale of difficulty, with respect to the rates of others of the same age. This is seen in the approximate formula used in the Stanford–Binet test for calculating IQ scores from the test, viz.

$$IQ = \frac{\text{mental age}}{\text{chronological age}} \times 100$$

Here 'mental age' is the score calculated from the test, reflecting how far up the scale of difficulty any one individual has got. When an individual performs on the test so that his 'mental age' is equal to his chronological age, then he will have an average IQ equal to the average for his chronological age group. Numerically the ratio of his 'mental age' to his chronological age is then 1 and the corresponding IQ score is 100.

An IQ score thus means rate of change up the constructed scale of difficulty. Fast developers on this scale obtain high IQ scores, and slow developers low ones. In as much as IQ is intended as a measure of intelligence, the tests appear to be measuring intelligence by rate of development. How appropriate is this to our rough notions of intelligence? 'Rate of development' is certainly not an idea that enters into any of the attempts of psychologists to define intelligence. These attempts all refer to characteristics of cognitive ability at any one point in time. The extent to which an individual's behaviour shows these characteristics is, however, assumed to be roughly invariant over time, in that intelligence is assumed to be a stable characteristic of an individual. This definition of IQ as rate of development (in as much as IQ is considered a measure of intelligence) means that we can no longer say that someone is a slow or late developer but also an intelligent person. Such a definition means we have to abandon the distinction between slow development and lack of intelligence. This is very unfortunate because, although many individuals who develop more slowly than average will also be regarded in many respects as unintelligent, this is by no means always so. The 'late developer' is a well-recognized phenomenon, and we do not necessarily mean by this someone whose intelligence changes, who starts by being relatively unintelligent and whose intelligence increases.

We must now consider in a little more detail how IQ scores

are calculated as there are several further points that have a bearing on their interpretation.

Firstly, an IQ score is the sum of scores on the great variety of items that make up any one test. In calculating an individual's 'mental age' on the Stanford–Binet test, the number of items at each level that are passed are simply added up. This means that the same mental-age score can be obtained in many different ways – in principle two individuals could obtain the same mental age score but not pass any of the same items, beyond a certain minimum level. This is so because all that matters is the number of items passed. In practice individuals of any one mental age do vary considerably in which items they pass, and which they fail, but there is also some overlap. Thus the same mental-age score, and hence the same IQ, may hide considerable and often extremely important differences in the pattern of cognitive abilities between individuals. The Wechsler tests go some way to recognizing this by separating the total IQ score into two parts – performance IQ and verbal IQ, but each of these is an amalgam of several diverse abilities. The new British intelligence test avoids this averaging difficulty to some extent by producing a profile for each individual for the several sections of the test, instead of one number. This profile shows the relative contributions that each subsection of the test makes to the final score.

A second point is that tests are so constructed that they give what is known as a 'normal distribution' of IQ scores. This means that the scores are arranged so that as many people obtain scores above the average as below. A graph of IQ scores plotted against the number of people obtaining any one score gives a bell-shaped curve (see figure 5 in chapter 5). IQ tests are constructed, and the raw scores are so transformed, that IQ scores obtained from any large and representative sample of the population will approximate to this distribution. This does not mean that what the tests are intended to measure, i.e. intelligence, is distributed normally or any other way. The normal distribution of IQ scores is simply a consequence of certain procedures used in test construction and standardization in order to obtain this end. The average IQ is designated as 100 purely for convenience, and it has no particular significance. Further, the normal distribution of IQ scores does not mean, as it is often argued, that IQ is determined by a random or chance combination of genes. There is no biological necessity in statistical artefacts.

Standardization of tests

We have described how IQ scores are an expression of an individual's performance on the scale of difficulty relative to his age group, but we have not so far discussed how norms are obtained for each age group. Obtaining such norms, and thus ensuring that the test fairly represents the abilities of each age group, is a major part of test construction, and it is what is known as standardization. Standardization is extremely important since the IQ score expresses an individual's relative standing, compared to all others of the same age. The test constructor has to know what is normal for any one age, to provide the scale for calculating IQs. This is done by trying out the test items on as large and representative a sample as possible, in the hope that this sample will resemble the populations on which the completed test will be used. The nature of the standardization sample effectively determines the future usefulness and applicability of the test. This is seen clearly in the cross-cultural limitations of tests. It is irrelevant and meaningless to apply American tests of intelligence to totally different cultures, since almost all the factors known to determine performance on a test (let alone the unknown ones) will vary between the two societies. However, even within one society there are reasons for supposing that the standardization procedures are inadequate and biased, despite heroic efforts on the part of the test constructors. Both the Stanford–Binet and Wechsler scales were originally standardized in America. Care was taken to obtain samples of children who conformed in number to the proportion in the population included in the latest census, with respect to certain variables, including urban or rural residence, socio-economic status, and education of parents. In the case of the Stanford–Binet the standardization sample came out as above the average for the whole population in socio-economic status (Terman and Merrill, 1960). This bias is likely to be further increased by the fact that the test constructors took only the population included in the census as their reference group.

They did not include the large number of migrant and unemployed workers who tend to be lost from a census. Further, the Stanford–Binet and the Wechsler scales were standardized on whites only with no explanation about this on the part of the authors.

These biases in the standardization of tests have several consequences. Firstly, when the tests are applied to populations not so selectively biased in the direction of superior economic status, the distribution of IQ scores will tend to be different from that of the standardization sample. Since IQ and socio-economic status are positively correlated, the IQ scores of the population in general will be lower than the average of the standardization sample, that is, lower than 100. This means that more than half of any general population tested are likely to have IQs below 100, and this is what is usually found. Some people have used this to claim that there is a greater incidence of subnormality than theoretically there should be, as predicted from the normal distribution of IQ scores. However, all it shows is that the standardization sample was not truly representative. The second consequence that follows from the white-only standardization samples is that these tests are only tests of white abilities. The tests are constructed on the basis of what white Americans can do at particular ages. Therefore, in as much as they are considered tests of intelligence, they do not and cannot measure the intelligence of black Americans. To do this, new tests, standardized on blacks only, would have to be constructed. We can still compare black and white children on the existing IQ tests. However in doing this, we are not comparing black and white intelligence, but instead how blacks do on tests of white intelligence (see also chapter 3).

It is probably impossible to standardize a test perfectly and without bias. This would not matter if it were explicitly recognized that a test is only applicable to populations resembling the standardization sample in all relevant respects. This limitation on the applicability of tests is not sufficiently appreciated, especially by those who attempt to make generalizations about interracial differences in intelligence on the basis of tests constructed by and standardized on one race only.

Conclusions

We have seen that an IQ score expresses an individual's rate of change up a scale of difficulty, relative to others of the same age group. This scale of difficulty is constructed according to certain statistical criteria, outlined above, in order to ensure that the test has certain desirable properties as a measuring instrument. No theories or findings about the development of intelligence in children are used in selecting items for this scale. The test as a whole is usually validated, if at all, against the external criterion of school performance. It therefore comes as no surprise to find that IQ scores do in fact correlate highly with educational success. IQ scores are also found to correlate positively with socio-economic status, those in the upper social classes tending to have the highest IQs. Since social class, and all that this implies, is both an important determinant and also an important consequence of educational performance, this association is to be expected. This chapter has shown that IQ is not, and could not be, a measure of cognitive abilities abstracted from all social and motivational factors. In as much as IQ tests measure anything, they measure the likelihood of educational and social success in a particular society. This is not to deny that cognitive abilities do contribute to such success, but rather to claim that it is impossible to consider such abilities in isolation from their social determination and expression. The assumption on the part of intelligence-test constructors that this is possible, combined with their preoccupation with the technical details of test construction, has given the concept of IQ a quite spurious aura of scientific respectability.

Further reading

A. Anastasi, *Psychological Testing*, 3rd edition, Macmillan Co., 1968.

H. J. Butcher, *Human Intelligence: Its Nature and Assessment*, Methuen, 1968. Currently the most comprehensive and detailed account of intelligence testing.

J. W. B. Douglas, *The Home and The School*, McGibbon & Kee, 1964.

A. W. Heim, *The Appraisal of Intelligence*, Methuen, 1954. This includes a critical description of the complexities of test construction.

D. A. Pidgeon, *Expectation and Pupil Performance*, Almquist & Wiksell, 1970.

L. M. Terman and M. A. Merrill, *Stanford–Binet Intelligence Scale*, Houghton Mifflin, 1960. This is the manual for the administration of the test: it also provides information about the construction and standardization of the test.

3 Can Racial Discrimination Affect IQ?

Peter Watson

Peter Watson is twenty-eight and was educated at the universities of Durham, London and Rome. After postgraduate work, he was a member of the clinical staff at the Tavistock Clinic before going to the Institute of Race Relations to carry out research into the intellectual development of immigrant children and to found and edit the institute's monthly, *Race Today*. He then joined *New Society* where he is now Associate Editor, but still finds time to carry out a certain amount of psychological research.

People are not as nice as psychologists say they are. We are less intelligent, more conventional, less sociable, more senile, less well educated and, very likely, more maladjusted than they have led us to believe. How do we know? Because recently some of them had the belated idea of looking at what kind of person it is who volunteers to be a subject in pyschological experiments – the sort of thing that a lot of our knowledge about behaviour is based on. And what the psychologists found was that these volunteers tend to be – among other things – better educated, more intelligent, better adjusted, younger, more sociable and less conventional than individuals like you or me who never get invited near a psychologist's lab.

Not that the psychologist's subject – his workmate, if you like – is the only thing to have turned out to be less reliable than was previously thought. The workman himself has some explaining to do. Several times recently it has been urged that psychologists should declare their backgrounds and their beliefs about behaviour when presenting the results of their own experiments. This is because those beliefs and backgrounds can be shown not merely to influence what a psychologist studies, but to regularly affect the sorts of results he gets as well.

But above all it's the psychologist's tools that have taken the worst knocking. Interviews, questionnaires, personality tests, intelligence tests: all these, critics say, are unreliable, producing a variety of different results depending on the conditions under which they are used. Give a child an intelligence test immediately after one kind of lesson, for example, and the score he gets will be different from what he would get if he took the test after a lesson of a totally different kind. Vary the seating arrangements for the test, or put him in with others instead of on his own, and you get still more discrepancies. Joanna Ryan's chapter brings out in more detail the differences of view, even between psychologists, on what intelligence tests are, what exactly they measure, and in what circumstances they should or should not be used.

Undoubtedly, the critics have at least part of a case – but many of them seem to have been just a bit too over-eager to label all the methods of the psychologist as unreliable. And in doing so they have overlooked the fact that, precisely because the psychologist's tests *are* influenced by conditions, they are, for that very reason, sensitive to them. In fact, because of this the critics have let a few psychologists show them a clean pair of heels – for quite a few tests are now used, not for layering people along some particular ability, but for systematically studying those factors which influence the abilities these tests measure.

So far as intelligence tests are concerned, you can see from the rest of this book that psychologists are by no means agreed among themselves as to what it is that these particular tools measure. But there can hardly be any disagreement as to what the tests themselves are. We ought, therefore, to able to agree at least on what can influence performance on these tests. Whether or not we eventually decide that the tests do measure intelligence, this approach should give us a working idea of the kind of influences we can expect, how they operate and what their extent might be.

It is certainly true that, quite apart from factors like nutritional deficiency or inadequate childrearing that affect intellectual *development* and are discussed in other chapters, almost every conceivable kind of variation in the conditions under which a test is given can affect someone's performance on it. Even the time of day seems important for some. In another case, nine and ten year olds were given an IQ test straight after they had written an essay in their English lesson entitled, in the case of half the children, 'The best thing that ever happened to me' and, in the other half, 'The worst thing that ever happened to me'. The children who had been writing about the less cheerful things scored, on average, four or five points lower on the test than did the other half.

But one of the most intriguing influences that psychologists have come up with (intriguing because it is unexpected), is that the individual characteristics of the person who is actually *giving* the test can affect how well someone does on it. These characteristics may be bits of the tester's behaviour which, once he is aware of them, he can control. For instance, if the tester is aloof and rigid in his manner, people score lower than when he is more natural, warmer. Other results show that if he expects someone to do either particularly well or really badly, that person, so it seems, is inclined to oblige.

On the other hand, there are some characteristics of the tester that appear important but which he can do nothing to avoid. He can't, for instance, do anything to alter his age, his sex, his class (not in the test, anyway) or his race: yet all these, we now know, influence the performance of the people taking the tests he gives.

Now, why should it be that a tester's age or his race – or any of these characteristics – have the slightest effect? You could argue that the tester is likely to be more of a stranger to the person taking the test if he comes from a different background, and that he is consequently likely to be more upsetting and off-putting. But this can't be entirely true because in some cases, at least, the person taking the test actually does better, not worse, with this kind of stranger as tester.

The line that a number of psychologists have followed in this is one that the famous psychoanalyst, Erik Erikson, has put well. What he said, in effect, was that the testing situation should be seen as a microcosm of society, that the particular face-to-face contact you get in a test situation, especially where the tester and person taking the test belong to separate groups because they have different backgrounds, will be affected by the prevailing relations between their two groups in the wider society. In other words, if the race of the tester – or the sex, class or age – influences how well someone does on the test – as it seems to – this is because the generation gap, the battle of the sexes, class conflict or the race war (whatever it is you call these relations) have wide-ranging psychological consequences on the individuals in the separate groups, one of which is to affect intellectual efficiency.

If this is true, it should follow that, by systematically changing the tester, and observing the changes in performance, we can get some idea of the way abilities are influenced by wider social forces. The influence of the race of the tester, especially in America, has been studied much more than the influence of any other group – possibly because someone's race, on the whole, is one of the easier things to get across in an experiment, but also, of course, because race differences in ability, rather than age differences, or sex differences, have become such a sensitive issue.

As early as 1936 it was known that when the IQ of American Negroes was tested by white and Negro testers, the Negroes scored, on average, six points lower when tested by whites than by Negroes. But not until the 1960s did the significance of this result sink in – and even then only to a handful of psychologists. Indeed, it may only have happened because the demands of civil rights groups raised once more in people's minds the question of how ability does or does not relate to success in life. However, between 1936 and the 1960s at least two other studies had showed results strikingly similar to this earlier one. There was, for example, a study in which American Negroes were asked to identify a number of famous names: they got more right when they were asked the questions by another Negro than by a white. And, in a later experiment, Negroes were asked to give synonyms for certain words (like 'emanate' and 'space'). Here too, they got more right if the questioner was a Negro.

With all this evidence, a few psychologists at last came to think that the fact that the race of the tester made a difference, to some people at least, was more than just an intriguing technical problem to be overcome. They began to think that tests should indeed be seen in relation to a wider context, as Erikson hinted.

For a time, though, this line was obscured by a number of others in this general field. General trends in migration, especially after the Second World War, boosted psychologists' interest in the problems of assessing the abilities of the various racial groups around the world. Reasonably swiftly, a good understanding was developed of the variation in psychological characteristics of groups as varied as Canadian Indians and Eskimoes, certain African tribes, Australasian aborigines and West Indians. Anything that could be expected to affect how well a person performed in his own culture and which might affect how well he did on I Q tests (like perception, language, thinking habits), were studied in detail. In this way, a technique was developed by which, within reason, it was possible to test a wide range of racial groups. Research showed how the standard western procedure should be changed at critical points – when to use, for example, three-dimensional material instead of two, when to change the presentation of instructions, how to vary timing. It was found that this manipulation upped the scores of various groups by six or so points and that tests became much more reliable – you got similar results when you retested the same people.

It was possible to use these adapted tests to predict, reasonably well, the school or job success of these groups in their own setting.

What, however, this concentration on technique masked, was that once different racial groups lived in societies with other, more powerful groups, another force came into play. Quite apart from someone's unfamiliarity with the test, quite apart from the fact that he might have a background that, nutritionally or environmentally, handicapped him, he would now have to live in a situation where he came into regular contact with people of a different race, and, given human nature, this could be stressful.

So in the early 1960s psychologists became aware that the fact that a black person scores higher on a test when it is given him by a black reveals not just a technical perversion but that for him a test with a white tester may be stressful. In fact, by 1960 quite a bit was known, not just about the effects of stress on behaviour but also about the way stress affects intelligence. The general conclusion of a fair number of studies was that stress actually improves performance on relatively simple tasks but hampers it, sometimes severely, on difficult ones. This reasoning immediately suggests a reinterpretation of the conventional view of race differences in IQ. At that time, this view was substantially the same as it is now – that in the case of black and white Americans, 1. whites regularly score about 15 points more than blacks; and 2. Negroes do better on simple tasks, like rote memory, than on more complex tasks, like verbal reasoning. This could mean, as Arthur Jensen was to argue notoriously ten years later, that blacks in fact were not only less intelligent than whites but that this intelligence was also slightly different in kind . . . more rote-like, less abstract. On the other hand (and, except for a very few psychologists, this went without notice), the lower score could mean that blacks were indeed acting under stress, for the shape of their performance, better on the simple rote tasks, poorer on the more complex reasoning ones, is just what theories about stress would predict.

The picture was building up, then. Still more pieces fell into place. Till that time, it had been a matter only of general concern and moralizing that friendships in multi-racial schools were formed usually along racial lines. No one thought it could have anything to do with ability differences between black and white pupils. Yet, when one study of race relations in a children's summer camp showed that whereas whites readily expressed the hostility they felt towards blacks (by fighting, for example), blacks repressed their hostility (and turned it inwards in the form of wetting their beds or nightmares), the red light was clear. A good number of experimental studies had shown that people who have difficulty in expressing their aggression openly suffer intellectually (this is true whatever their race). In one of these studies, two groups of students were given a projective test and divided into those who were and were not unwilling to give aggressive responses. Then, on a later test of anagram solving, it was found the inhibited group did less well.

You can see the picture emerging: that even the general atmosphere of race relations can affect the performance of the groups on the receiving end of hostility and discrimination. Something more, however, in the way of experimental analysis was called for. Two simple laboratory experiments confirmed, though, that the reasoning was in the right direction.

For example, Irwin Katz, a psychologist working in New York and the mid-west, devised a neat way of testing not only whether a white environment for a black is stressful, but also whether IQ tests are, too.

He gave Negroes a problem to do, either in the presence of two whites or else with two blacks. One of the two posed as another person doing the same problem; the other played the part of the tester. Stress was added in each situation in such a way that it could be controlled – the Negroes were warned that they could expect, during the problem, either a mild or a severe electric shock. Their performances varied according to the particular variety of tester and type of shock. Where they expected a mild shock the Negroes did better with the white tester; when they were warned it was to be severe they did better with a black. This is what stress theory would predict. Mild stress (mild shock plus white tester) improved performance. But the combination of severe shock plus white tester was too much and performance dropped. The Negroes did quite well with a black tester and severe shock. I should add, of course, that no shocks were ever given.

This was encouraging, so the next step was to see whether IQ tests were also seen as 'stressful'. The pattern of the previous experiment was therefore repeated exactly, except that instead of being told they would be mildly or severely shocked, the Negroes were told they would be given either a research exercise to do or else an intelligence test. Identical results were obtained: blacks did much worse on the problem when they were told it was an IQ test and when it was given by a white person, than in other conditions – even though the task they had to do was exactly the same in all conditions.

Katz also gave a number of Negroes a different kind of test. This was in appearance a concept formation test. There were about sixty items, each consisting of four words. The odd word out, in each four, had to be circled to get a correct answer. However, in some cases there were two possible right answers, except that one involved choosing an aggressive concept, whereas the other didn't. (An example might be: Belt; shoe; sock; hit. Here, if someone chose 'shoe' as the odd man out, the concept left would be aggressive, whereas if he chose 'hit' the concept left would be 'clothes'.)

Katz gave this test to Negroes describing it either as an IQ test or as a research tool he was trying out designed to see how people learn to use words, emphasizing in this case that it had no connection with ability. Furthermore, the people who explained this to the students were themselves either Negroes or white. The most striking thing about this procedure was that when the test was given by a Negro *and* described as an intelligence test, the aggressive scores of the Negroes went right down – in other words, if they had strong feelings they appeared to repress them.

The experiments, therefore, supported the general observations: when a Negro takes a test with a white tester there may well be stress of a kind which affects his performance. The white person is seen as a threat and evokes an emotional reaction in the Negro. But the analysis can go further. Look again at the situation from the black's viewpoint. He has migrated, possibly, from a less developed country to a more developed one, or from the deep south to the north. But whatever the actual circumstances, they are likely to have one thing in common: so far as ability is concerned, the black will now be moving in a world where standards are higher. His chances of success will now be very different from what they were before. Psychologists found that this latter idea could be useful in explaining differences in ability.

For we know that the *expectations* a person has about how well he will perform some task actually influence how well he does perform. On the whole, it has been found that when someone's chances of success are around 50:50 his motivation to succeed is at its highest. As chances get better, he loses interest; as they get worse, he loses hope.

How then does this fit into the black's world? Everything clearly depends on how good he thinks his chances are. And there is a way that this can be controlled and assessed.

Katz at first looked at this problem by giving Negro students a test and telling them that they would be compared on the results either with other Negroes or with whites. (He did this indirectly, to make it realistic, by saying that the students would be compared either with other students in their all-black college or, on the other hand, with all the students in the mainly white state.) The Negroes scored significantly better when they anticipated comparison with other Negroes than with whites.

Strictly speaking, though, this experiment does not entitle us to say that low expectancy of success was entirely to blame. Simply being compared with whites could have been stressful. So in a later experiment Katz controlled the level of expectancy more rigidly. Since it is the crucial experiment in this field, I will describe it in some detail.

Negro students at a college in the U S deep south – which admitted only Negroes – were given a particular task by a Negro. The task involved learning to substitute different shapes for numbers and its purpose was to get a level of ability for each student. Later on, only *changes* in the level of performance were recorded. A few days later, the students were again tested with a different version of the same test – this time in small groups. The tester now was either white or Negro and explained that he was a psychologist from the 'Southern Educational Testing Service' – an organization which, in practice, does not exist. He told the students that the earlier testing session had, in fact, been only a tryout for a 'scholastic aptitude' test that they were just about to take. They were also told that norms for this test had been worked out – so comparisons of their performance would be made either with those for freshmen of their own college (i.e. black) or for the state overall (white). To heighten the realism, they were told that immediately after the test was taken, the psychologist would see each person individually, mark his paper and discuss each student's aptitude with him. Finally, and just before the test was given, the psychologist handed to each student an envelope with each person's name typed individually on it. This letter was ostensibly from the 'testing service' and explained that from the subject's score on the practice trial it was possible to calculate how likely it was that he would get the average for his age group on the test. In practice, of course, the probabilities that were typed into the letter had nothing to do with any previous performance – but the figures that were typed in were either 10 per cent, 60 per cent or 90 per cent. A control group were not given any letters at all. Finally, the students did the test.

In all cases the biggest gains over performance on the trial test were achieved where students were told in their letters to expect 60 per cent success. Secondly, the best gains of all were made by students who were told to expect this 60 per cent success by a white tester and expected to be compared with other Negroes. What seemed to happen here was that blacks who knew they would probably do well, and having no anxiety about being compared with whites, relished excelling in front of a white. In a secure situation for blacks, there could well be more incentive to do well in front of whites since their approval would be worth more.

On the other hand (and thirdly) when the test was given by a white tester and white norms were the comparison, performance of the students was very bad. Finally, when they were reminded, on top of this, that they shouldn't expect to do very well (10 per cent probability) performance was even lower. But, with a white tester and white norms for comparison, performance under *no* specified feedback was closer to that when a 10 per cent probability of success was to be expected than with a 60 per cent probability. This would imply that when they were not given any idea of how well they were expected to do, Negroes 'naturally' thought that, compared with whites, their chances of success were very poor indeed.

It would seem, therefore, that there is at least a double handicap to your intellectual performance when you are black: 1. the white environment, particularly in America, is threatening and stressful, evoking reactions that are a drain on your performance; 2. your expectancy of success is low (realistically, usually) and this only makes matters worse. In testing disadvantaged groups, much less attention has been paid to these handicaps than to those arising from differences in upbringing or, in the case of immigrants, from language problems. But there is still more evidence that suggests the factors I have mentioned are even more important than I've been able to show in this short review.

In another study, for example, each of the whites and Negroes who performed equally on a test were asked to rate how well they thought they themselves had done. Negroes, it turned out, rated themselves lower – that is, *unrealistically* lower – than the whites rated themselves. So as well as having real drains on their ability, it looks as though some Negroes have imaginary ones as well – stemming perhaps from the way they have assimilated popular ideas about themselves and their ability.

Consider that experiment where 'probability of success' was manipulated. In that experiment, reminding Negroes of the white world outside affected their performance. One of the things that you could deduce from this is that if a mere *reminder* of the outside world in a test situation provokes such marked deterioration in performance, what must a lifetime of being discriminated against, having to pit yourself against higher standards, do to your intellect? The effects produced in these experiments may be only a few IQ points – but over a lifetime they could be much more.

Above all, what these studies show is how important it is to look at a situation from the point of view of the person who is in it. Surprising differences occur that way – as well as possible unexpected similarities in what appear to be entirely different contexts. Race relations in Britain, for example, may look very different from the US situation. But that doesn't stop some features being similar.

In 1968, I gave a schoolful of West Indian teenagers a concept formation test, with either a black tester or a white one. Sure enough, however nice we think race relations are in Britain, the children tested by a black tester scored several points higher than those tested by the white man – me. Three years later, Jill Hodges and I gave a scale very similar to the one Katz and I had previously used, this time to West Indian school children of different age groups in East London. Once more, the scale was given either by a black or a white. Our results differed in some ways from Katz's – as you would expect – but were nevertheless similar in other important ways. In our fourteen to fifteen-year-old group, the combination of a white person administering the scale and giving 'test' instructions produced a sharp fall-off in performance. With a black administrator and 'test' instructions, or with a white administrator and 'no-test' instructions performance went *up*, as in some of Katz's experiments. With seven to eight year olds, however, a white tester made performance worse, no matter what the instructions (though the fall-off wasn't as great, at this age, as it was later on in white administrator/'test' instructions). And in all cases, at all ages, when performance went down, aggression scores went up.

One may, in all these experiments, begin to have some idea of the way 'race' and 'ability' get tied up in the mind of the youngster and how this, in turn, can both improve and impair his performance.

We don't know yet just what detailed implications this has for the intellectual performance of West Indian children in Britain, though clearly there is likely to be some effect. This and many other questions must be answered, before a conclusive picture finally emerges. But there is enough evidence to enable one to say this: till now, psychologists, whatever their views on the origins of differences in IQ, have recognized only two kinds of environmental influence – those related to childrearing and those related to cultural differences. It is time a third was added – differences in motivation due to chronically poor race relations. If the results of the experiments are to be believed, and the ideas lurking on the horizon turn up trumps, this could be one of the strongest drains of all on ability.

Further reading

Martin Deutsch, Irwin Katz and Arthur Jensen (eds.), *Social Class, Race and Psychological Development*, Holt, Rinehart & Winston, 1968.

Thomas Pettigrew, *A Profile of the Negro American*, Van Nostrand, 1964.

Peter Watson, 'How race affects IQ', *New Society*, 16 July 1970.

4 Jensen, Eysenck and the Eclipse of the Galton Paradigm

John Daniels and
Vincent Houghton

John Daniels is the son of a Staffordshire collier. He is now Head of the Further Professional Training Division, School of Education, University of Nottingham.

Vincent Houghton is Head of the Division of Behavioural Sciences at Huddersfield Polytechnic, and formerly in the behavioural science department of Simon Frazer University, Canada and at Nottingham University. He has studied a variety of children of different ethnic groups, and believes that basic respect for a different culture is essential before any scientific investigation can be meaningful.

The Black Papers on education produced their own special effects. Anger at provocative and unprovable assertions was matched by moments of light relief. For one contributor, however, H. J. Eysenck, the Black Papers were only rehearsals for a full-scale performance in *Race, Intelligence and Education.*

Eysenck spent his school days in Hitler's Berlin and, later, as a young man, came to England just prior to the war. Here he encountered the stupidities of the English system of meritocratic higher education; he found that because he had the 'wrong' pattern of subjects, he could not enter the university to study physics. As a substitute, he chose the less 'demanding' subject, psychology. Psychology had still not achieved full respectability as a university discipline, but it was reputed to be on the threshold. Sir Francis Galton had pointed the way forward for psychology; he had invented a new school or system of thinking about people which we shall refer to as psychometry. Psychometry was dazzling the world, demonstrating that it could penetrate the mysteries of humanity by applying complex multi-dimensional mathematical methods to the psychologists' 'hard' data – just like the physicists had conceived of atoms after Newton.

Psychometric hard data, to be sure, looked to most people who had attended school remarkably like the familiar, and fallible, examination marks, but, reassured the psychometrists, test marks are more 'objective' than the introspective judgements which most psychologists had worked with before Galton. Paper-and-pencil examinations, it was thought, could have their respectability restored by a new P R campaign and a new set of names. After all, the class-divided system of education which was, in the early part of this century, absolutely dominant still had need of examinations. What better if schoolmaster and psychological academic could link hands, one in self-justification and the other in search of respectability for his 'science'.

Psychometry, during the years of the Second World War, in Britain achieved new heights of power. Psychometry was built into the war-machine with an important psychological role to play – to keep the conscripted soldiers thinking that in the 'democratic army' everyone had a 'fair' deal. All were being given the jobs, officers or 'squaddies', which they were 'capable' of doing. Millions of men for the first time encountered the psychologist. Millions of pieces of the new 'hard data' were collected – intelligence test results, K-factor test results, Ink-blot interpretations, V-factor test results. Psychology had made it.

Intelligence testing after the war totally invaded the schools. The scholarship examination for entry into secondary schools was declared out-moded. It had to be replaced by the psychometrists' own refurbished version of the old general knowledge exam; the test of intelligence combined with the 'objective' maths test and the 'objective' English test.

During this psychometric field-day, the psychometric boffins, including Eysenck, were not idle. The dream of psychometry was not only to bring the mind (the cognitive processes) into mathematical review; why could not the same tools now be turned against the old fortress – human personality itself? Why could we not create 'tests' of personality? We could reduce personality to behavioural units (test data) and apply the tools of correlation, partial correlations, multiple correlation, factorial analysis. In this way, we could crack the code of human personality.

So psychometry identified factors of 'introversion–extroversion', 'radicalism–conservatism', 'neuroticism–stability' and a host of other self-generating and circuitous factors. The faculty chaos that Spearman complained of and which was ostensibly the reason he gave for developing factorial analytic methods in the cognitive sphere, had now led to the much more confusing chaos of personality factors. This mish-mash of eclectic thinking about human learning, human activity, human cooperation in work is the complete opposite of behavioural science and of any 'objective' approach to understanding of humanity.

In one sense, the rise of psychometry coincides with the rise of a number of related ideologies which have the function of rationalizing many popular strivings for a restructuring of society. Both feudal and technocratic kinds of job allocation were concerned with that kind of allocation which was based firmly on the needs of the ruling strata of our society. In the century 1870–1970 there were forces which pressed for greater democratic participation at all levels of society. These pressures had profound effects on psychometric reasoning.

The Galton paradigm is itself essentially anti-democratic. During its growth period, it was devoted to presenting reasons why various strata of society should be excluded from decision-making processes. This was particularly true in education. Fundamentally, psychometry sought to show that only a small minority of the population had sufficient '*g*' (general intelligence) to profit from professional and higher education. The method of intelligence-test construction itself automatically and circularly determined at the same time that *g* could be found in abundance only amongst upper and middle-class children. The sharp battles for comprehensive schools during the democratic advances which followed the Second World War threw psychometry into deep crisis. For example, psychometry had always talked about its need to give special help to 'duller' children inside the schools, i.e. providing classes for backward children, ESN schools, 'opportunity' classes, providing special educational help to those who were unfortunately lacking in *g*. This liberal chatter delayed the day when the structure of psychometry could be called into question as in itself anti-scientific. What had now to be questioned was the very concept of intelligence itself. Are there any scientific grounds for believing that there are such entities as 'factors of the mind' that explain in any useful way the behaviour of human beings? The answer now comes out clearly in the negative.

In America, as one would expect, psychometry had a parallel development with its own specifically American overtones. Intelligence tests had been used as early as 1917 in officer selection for the new American conscript army. As a by-product of the data these practices provided, there was much discussion on so-called racial differences in intelligence. Naturally, racialist elements were ready aplenty who seized upon this data to 'prove' that white Anglo-Saxon Protestant elements showed considerable superiority as measured by intelligence-test scores over the ex-slave black elements. Liberals and conservatives fought wordy battles concerning the 'meaning' of the data but by 1969, with the blacks bursting out of their ghettoes, the main leaders of psychometry thought it politic to announce that, in their view, there was no evidence to show that there are inborn racial differences in intelligence. All differences in mean scores could be explained in terms of environmental deprivation.

Jensen and Eysenck, in questioning this view, have clearly spelt out what they see as the implications of working within the Galton paradigm. They roundly declared that their calculations show that *whites* are superior in 'intelligence' to *blacks*. They have both replied at length to their critics in arguments about the meaning of the data they have examined, and have given at least as good as they have received. It would, however, be an exaggeration to grace this inter-change with the label of 'scientific debate' because, with both groups, the evidence was selected to prove a stance to which for quite other reasons, each group was already committed.

Social scientists have tended to have the same relationship with their data as government officials. In the words of Ossowski, 'The relation of the state to the sociologist is that of a drunk to a lamp-post, it wants support, not light.' One need not be a totally committed follower of R. D. Laing to see that the adjustment of the individual to a destructive and dead-end social situation is the obvious political aim of much psychotherapy. In psychometry, to secure acceptance of one's allotted place or role as given by the educational statisticians is a political act in the same way. The maintenance of the *status quo* is the central duty. This is the core of all the arguments within the Galton paradigm. That is why we owe such a debt to Jensen and Eysenck because they have clearly demonstrated the consequences of accepting the psychometric view of man.

Galton's psychometry gives a coherence to society which results in the appearance of justice in the allocation of social roles. In the Soviet Union, the meritocratic educational system has produced similar results by a straight examination system. There is essentially no difference between an examination system and a system based upon psychometric assessment. Both are used for social-role allocation; in the Soviet Union the examination system is better because a purely hereditary principle would have been ideologically unacceptable. Of the two systems, the one which is likely to last longer is the examination system because of its lack of pretence to a scientific rationale. If the demands of society change, then, theoretically, an examination system could change in quite an arbitrary way, untrammelled by assumptions which are not concerned or supported by so-called scientific notions of the 'properties of man'. Both methodologies, the examination system and psychometry, will undoubtedly be with us for some considerable time to come, and will only be replaced when another system of role allocation suggests itself as technocratic development reaches its point of explosion.

We have put forward the thesis that both examinations and psychometric tests have outlived their usefulness because, by their very nature, they cannot adapt to the changing needs of society. They play a key role in allocating pupils to special roles within society at the same time as they claim to objectivity, i.e. truth irrespective of the society in which they are used. Competitition for employment opportunities in our system of meritocratic job allocation as more and more people compete for fewer and fewer production jobs whilst production itself continues to soar, inevitably increases as the number of skilled (that is, top technical) jobs decreases. Temporarily, this could theoretically be mitigated by concentrating resources, for example, upon threats of war and preparation for it, or by diverting young people towards higher education. Both have been tried recently in the United States with conspicuous failure. Both have also been tried recently in the Soviet Union, and it is probably due to the slower rate of technological advance in that country that the failure there has not yet been so evident. However, the exclusion of a large number from the labour market by increased military or space expenditure or increased facilities in higher education only delays the date when the problems of what people are to do with their enforced leisure time become acute. There is, however, little evidence that the postponement will be profitably used in a positive way to meet the coming day of reckoning. Meanwhile, the nineteenth-century system of grading human potential will also continue. Professors of educational psychology will produce more and more specious rationalizations. The need is, however, for the wholesale restructuring of the role-allocation system, and this will be seen when it is clear that the professors are up against the stark issues of human survival.

This is the contemporary social background of the continued use of the Galton paradigm. Perhaps at this stage, a clearer definition of that paradigm would be helpful. In his preface to the original edition *Hereditary Genius* Galton is quite explicit.

> The idea of investigating the subject of hereditary genius occurred to me during the course of a purely ethnological enquiry into the mental peculiarities of different races, when the fact that characteristics cling to families was so frequently forced upon my notice.

The origin of this idea was a childhood dream to prove that his cousin, Charles Darwin, had not only, allegedly, found the clue to the origin of species, but in so doing, had also found the clue to the origin of different social roles, that is, social classes, within commercial society. Galton thought about his contemporaries at school, at college, and in later life. He then picked four hundred 'illustrious men of history' and 'the results were such, in my opinion, to establish the theory that genius was hereditary'. He wrote all this up in the book *Hereditary Genius*. Galtonian psychometry essentially begins at this point. Thus his theories are not based on any kind of inquiry which, in even the loosest meaning of the term, could be called scientific. If one produced today a list of four hundred names from ordinary family backgrounds who could reasonably be described as geniuses, it would be no answer to Galton because it would similarly be based on subjective criteria. It is more than likely that, given Galton's view of society, he was able to 'recognize' those who were obviously at the top of the social ladder and that this was a meaningful way of looking at Victorian society. Behind Galton stands the mechanical, schematic, atomistic way of thinking about phenomena which constitutes the Newtonian paradigm. In fact, Galton's way of thinking about society is within the Newtonian scheme. Behind Newton stands the giant of them all, Thomas Hobbes, who in *Leviathan* clearly anticipated Newton, Talcott Parsons, Darwin, Galton and Sir Cyril Burt. It is not difficult to understand why the mechanical balance model has lasted intact in psychometrically dominated educational psychology. The model, and especially Galton's version of it, of course denies the possibility of change. If intellectual functioning is fixed at the moment of conception, if 80 per cent of adult performance is directly dependent upon genetic inheritance, how have the styles of our lives and the patterns of our thinking changed to the extent that they have? Within psychometry, there is no answer to this question.

It is not necessary to study in any detail the application of the so-called Gaussian distribution of human ability. It is evident that amongst a highly evolved psychological species with a myriad of cultural heritages and a galaxy of qualitatively different modes of conceptualizations, any measures like intelligence-test marks, combined arithmetically, would inevitably produce something which was normally distributed. The emergence of the normal distribution is a function of the particular mathematical methodology employed, and nothing else. Nor need we here concern ourselves unduly with the anti-scientific absurdities of the correlation coefficient on which Professor Lancelot Hogben poured such scorn a decade ago in his *Statistical Theory*.

Since Gödel, the consideration of mathematical procedures as being culturally loaded artifacts, rather than a route to Aristotelian universals, is one which is nowadays widely accepted. Many research workers, even within psychometry, recognize that to choose a particular statistical procedure with which to handle a data is to choose the result to be obtained. The longer that disciples are able to deep-freeze a time-located theory, the more out-dated that theory becomes. That is why the pronouncements of Galtonian psychometry sound so archaic.

Galton had the misfortune to hand over his theories for development to the most brilliant social engineer of the early years of this century, Sir Cyril Burt. The unhappy dialectic between hereditary factors and environmental factors has been the result of Burt's superb ability as a propagandist. The answer to Burt's conundrum, how much is due to heredity and how much is due to environment, is not the point to fasten on. It is the question itself that must be questioned. The obsession with labelling people results from a genuine attempt to see diagnosis as a first step towards giving help to the casualties of our education system. But the question not only has no meaningful answer. It produces the disease which it is intended to cure. If, as psychometry contends, a child's limitations are the result of a poor genetic environment, little can consequently be done in the shape of remedial education. These remedial procedures, as Jensen correctly pointed out, have themselves failed and were doomed to failure, *if only because they were expected to fail*. They were expected to fail because of learned failure, because of the lack of helpful pedagogic strategy, because of nutritional factors, or because of poor genetic inheritance. The genetic factor is seen as one of a number of interacting factors in a complex situation. To overstress the genetic factor is inevitably to assert that the blame belongs on the shoulders of the victim. This whole method of diagnosis and treatment takes us into the theatre of the absurd.

In a sense, Eysenck is right to characterize himself in his book as an 'interactionist' when compared to the naïve environmentalism of writers like, say, those of the Iowa school of the 1930s. If the cat had kittens in the oven, you would not call them biscuits. Unfortunately, some writers, for example

J. B. Watson and the ensuing American environmentalists, fell into the trap prepared for them by accepting the nature–nurture proposition as spelt out by psychometry. They answer 'yes' to the question, 'Have you stopped beating your wife?' With genetic heritance, we must link the known importance of the period *in utero* catalogued in detail by Ashley-Montagu and the hazards in the process of conception itself described by geneticists in a recent article in the *Scientific American*. It is perhaps true that we are just becoming aware that the ratio of environmental factors which are easily susceptible to social manipulation is less than liberal voices claimed ten years ago. It would, however, be a foolish man indeed who would assert with confidence that a 4:1 ratio at the moment of conception is a ratio that we could take at all seriously.

The conflict and confusion on this matter is due, however, to an inability amongst social scientists to realize clearly that, as Heisenburg pointed out in physics, contradictions are the result of asking questions to which there are no meaningful answers. Eventually, social scientists will have to concentrate upon the relative consequence of various kinds of feedback in controlled learning situations. They will have to construct operational criteria which are open to modification. They will, therefore, have to answer questions of value, not in an absolute, fixed way, but in a way which is relative to the information they have available at the time. The backlash against pointless quantification and witless rigour, operating blindly, could even prevent the sensible utilization of the knowledge which has already been gained in this field. The denial of all structure and the belief in unstructured situations could prove to be as powerful and seductive a form of anti-intellectualism as domination by the theory of probability ever was. The Galtonian paradigm of a ladder view of society with the climbers moving with different weights given to them at the start of their journey preordaining the final result of the competition remains the dominant one, not only in education but in society at large.

This type of psychological Calvinism found support and sustenance in the latter part of the nineteenth century. The fact that it still finds overwhelming support testifies to the fact that the religious element in the Protestant ethic may have lost its appeal, but its commercial side has certainly not. Tawney's thesis might have been called 'Religion and the Rise of Industrialism' for it is the social feedback of technological advancement in the last two decades which has rendered social and psychological Calvinism obsolete. The fact is that vital educational arguments remain wedded to the ladder view of society and that the argument as to whether that ladder is narrow, wide or pyramidal, is seen as the important question.

The Galton paradigm, this ladder view of society, is now being applied by Jensen, Shuey and Eysenck to the races of the world. The assertion is that there is little possibility of change in an order of genetically determined levels of intellectual ability. Unlike the Football League, there is no possibility of promotion or relegation. How fortunate for those who have placed themselves in the First Division for all perpetuity! Galton was explicit about this. In *Hereditary Genius* (page 395), he writes:

> The number amongst Negroes of those who we should call half-witted is very large. Every book alluding to Negro servants in America is full of instances. I myself was much impressed by this fact during my travels in Africa. The mistakes the Negroes made in their own matters were so childish, stupid and simpleton-like as frequently make me ashamed of my own species. I do not think it any exaggeration to say that their 'C' is as low as our 'E', which will be a difference of two grades. [He adds,] The Australian type is at least one grade below the African Negro.

With 'scientific' caution, Galton adds, 'I possess a few serviceable data about the natural capacity of the Australian but not enough to induce me to invite the reader to consider them.' Since Galton, data has been amassed by many research workers. Based on the arbitrary selection of skills chosen by the white middle-class test authors, the Galtonian suspicion about Negroes and aboriginal Australians has now been largely 'substantiated'.

Wild assertions are made that the future of both society and individuals can be predicted by the relationship of individual test scores to mean scores. No doubt the history of the Lancashire cotton industry, the British shipbuilding industry, and the Yorkshire textile industry would have been different if decisions had not been made on 'average performance'. The arithmetic mean is one measurement which in a rapidly changing technological environment cannot be used to predict the future. There are certainly industries and social situations in which means provide a reliable short-term guide, but even then, uncritical acceptance of the argument from means is fraught with danger.

We have shown how important the contributions of Jensen and Eysenck have been to our understanding of psychometric psychology. By their rigid consistency and their careful use to selected data, they have shown that if one accepts the Galton paradigm as applicable to one's own society, then there is a very strong case for accepting it for any society or for any specific group of people.

In North America and the United Kingdom, perhaps in every advanced country, we are now educating the majority for unemployment. In North America at the present time, the unemployment rate of Ph.Ds under thirty-five is extraordinarily high. Perhaps in these new circumstances, the whole grading system, the whole scheme of meritocratic badges, is coming into question. The system of graded meritocratic assignment of social roles was never accepted uncritically in business and industry because it was known that those organizations which were most rigid and hierarchical most frequently went bankrupt. It is hard to see how the grading system can be retained when a guarantee of job opportunity can no longer be given. In a mad moment, we can just imagine a regime in which differing rates of unemployment benefit could be paid to those with different qualifications, top rates of pay going to those with Ph.Ds in theoretical physics.

Those usually in top jobs in education and administration often defend the grading system, sarcastically asking, 'Would you approve of amateur surgery?' We are not suggesting any such thing. Surgery, in common with a large number of other high-order skills, will always require *proven* competence. But this is no argument to justify the vast body of irrelevant examination organizations at every level or the widespread use of verbal intelligence tests. To a large extent, they are both tests of rote memory unrelated to specific behavioural skills. In a world in which vastly greater bodies of information are more readily available by other means, rote memory is a skill of declining importance.

The reasons for the continuation of the Galton paradigm as a phenomenon of functional autonomy lie elsewhere. We once heard a very distinguished educational psychologist referring to his colleagues of the 1940s and 1950s as having served their masters well. We have heard colleagues complain of the 'Professors of arithmetic' who have dominated British educational policy over the last three decades. While this kind of negative reaction to our current situation is understandable, it is less than helpful. In *The Structure of Scientific Revolutions*, Kuhn points out that previous knowledge is usually incorporated into the new paradigm. Examples of this are legion from the change from the physics of Newton to the physics of Einstein, Darwin's debt to Lamarck, perhaps even Kuhn's own incorporation of the style of thinking in the later works of John Dewey or to the towering intellect of Charles S. Pierce. A new comprehensive matrix emerges from the intersection of separate or even conflicting matrices of thought. Gradually it is itself submerged into a new and more appropriate arrangement. In education, we can see the need for greater flexibility in learning situations which perhaps will have to prepare students to change their occupation and, even their life styles, at least twice or three times in their lifetime. We are witnessing the break-up of traditionally encapsulated 'subject' disciplines and the emergence of a behavioural science using key concepts from a number of other related disciplines. In this new social scene, Galtonian modes of thinking about man and his roles in society will find only a little to contribute. Jensen and Eysenck, busy as they are, are really only busy in the propaganda of deploring the future.

Further reading

A. Anastasi, *Psychological Testing*, 3rd edn, Macmillan Co., 1968.

N. Chomsky, 'The case against B. F. Skinner', *New York Review of Books*, vol. 17, no. 11, 1971.

W. Heisenberg, *Beyond Physics*, Harper & Row, 1962.

L. Hogben, *Nature and Nurture*, Allen & Unwin, 1939. A classic analysis from statistical genetics.

L. Hogben, *Statistical Theory*, rev. edn, Norton, 1968.

V. Houghton, 'Intelligence testing of West Indian and English children', *Race*, October 1966.

J. McV. Hunt, *Intelligence and Experience*, Ronald Press, 1961.

T. Kuhn, *The Structure of Scientific Revolutions*, University of Chicago Press, 1971.

N. Miller and P. Dreger, 'Comparative psychological studies of Negroes and whites in the United States', *Psychological Bulletin*, vol. 57, pp. 361–402, 1960.

H. M. Segall, D. T. Campbell and M. J. Herskovits, *The Influence of Culture upon Visual Perception*, Bobbs-Merrill, 1966.

P. E. Vernon, *Intelligence and Cultural Environment*, Methuen, 1969. How the mind develops as long as education continues.

From Biology

In part 2 the biologists take up the discussion and examine the concepts of heredity and race. As Walter Bodmer points out in chapter 5, the biological concept of race does not coincide with the social definition. The social definition depends on where a person places himself and where the rest of society puts him, while the biological definition uses genetic criteria. He goes on to examine this biological definition and the types of evidence upon which claims for a genetic basis of race and class differences are based. Rejecting the old typological view of races as separate fixed types, he finds evidence for differences in gene frequency between the groups designated as black and white. However, there is no evidence to associate any of these differences with IQ scores, and as he shows, there are no *a priori* reasons for thinking that such a connection should exist. This view from genetics, far from asserting that all people are the same, argues for a high degree of diversity, a diversity arising in part from genotypic differences. This theme is one of the main points of John Hambley's discussion in the next chapter. He regards the infinite variety found among individuals as a resource to be valued and explains how it arises during development. This is probably the most technical chapter of the book but it is necessary to go into some quite complicated argument to explain why we cannot regard development as a simple putting together of genes and environments. Indeed in many ways this is the key to the whole discussion; so long as we accept the typology of nature and nurture we are led into false and misleading blind alleys. As John Hambley explains, development involves *interaction* of genes and environment at all levels. To try to take apart this interlocking system is like asking whether the breadth or height of a rectangle contribute more to its area. Another central point arises from this discussion – the fact that the genes cannot determine anything. Genetic factors are involved in the development of all our behaviour but because of the interactions during development there is an essential indeterminancy between genotype (the genes) and the phenotypes (adult characteristics as we see them). To predict adult characteristics from the genotype involves complete knowledge of the environment and of all possible interactions, and this is something we are beginning to see only in vaguest outline.

The last biologist, Steven Rose, explains the neural basis of intelligent behaviour. He describes some of the relevant evolutionary history and the development of brain function during an individual's lifetime. He points out some of the many ways in which environmental factors may influence this development and shows how such forces may act from generation to generation, and so give the superficial impression of being hereditary. One such factor he discusses at length is nutrition; this is clearly central to the main theme of the book as we have overwhelming evidence that minority groups like the blacks always tend to be less well fed than the majority. This may have crippling effects on cognitive capacity – a point to which we will return in the conclusion.

5 Race and IQ: The Genetic Background

W. F. Bodmer

Walter Bodmer is Professor of Genetics at Oxford University. Before this he was at Stanford University School of Medicine. He took his Ph.D in statistics and population genetics at Cambridge in 1959, after gaining a degree in mathematics there.

This chapter is a review of the meanings of race and IQ; and the approaches for determining the extent to which IQ is inherited. The first question to consider is 'What is race?' and then one must demonstrate how one can study the biological inheritance of mental ability as measured by IQ tests. These two aspects of genetics form the main underlying theme for this chapter.

What is race?

In almost all the psychological studies carried out on racial groups, race is defined sociologically or culturally and not biologically. However, biological race boundaries often coincide with those that are culturally evident, though not always. An example of a sociological or cultural definition of race that is not strictly valid biologically is that of children of black-white marriages in the USA, who are still regarded as black.

To a biologist a race is just a group of individuals or populations which form a recognizable sub-division of the species. The group is identified by the fact that the individuals within it share characteristics which distinguish them from other sub-groups of the species. The species itself is most simply defined as that set of individuals which includes all those who could produce fertile offspring if they mated with each other. This means, of course, that matings between members of different races are just as fertile as matings within races. The members of a race or a sub-group are however most likely to find their mates within their own group. This results in the groups becoming separated from one another as far as reproduction is concerned. Because such groups mate among themselves rather than with other groups, they tend, as we shall see, to become more and more different. The more different they become and the more they are separated from each other, the less likely it becomes that individuals marry outside their own group. This tends to make the groups become more and more distinguishable from each other.

The sorts of things that keep groups physically apart are, for example, mountain ranges, wide rivers, seas and deserts, and just distance alone. It took many thousands of years before Europeans crossed the Atlantic in significant numbers and came into contact with the American Indians. And even then, they hardly intermarried – it was mainly germs causing diseases like measles, which crossed the racial barriers. But even in a comparatively small country like England, distances are such that at least until quite recently, it was not very likely if you came from the north you would marry someone from the south. In fact, you would be most likely to marry someone from your own parish, or at least one quite nearby. Modern transport, of course, has scaled down the significance of distance as an isolating factor. Partly as a result of this, many of the differences that have accumulated in past years, even between neighbouring towns, are now rapidly disappearing. But most of the sub-divisions of man which are nowadays called races, originated long before the Industrial Revolution and modern transport. Greater mobility, as well as affecting local marriage patterns, certainly has an enormous impact on the larger sub-divisions of the human species by bringing together groups of people who had previously lived more or less in isolation from each other. This happened when slaves were taken over to the Americas from Africa and more recently when American blacks moved into the northern states and their cities, from the southern states.

Major migrations have led also in the past, often in the form of invasions, to new marriage patterns and the emergence of new population groups. The bringing together of different peoples should eventually bring about a blending of the differences between them, but at the same time it can be a cause of many of the racial tensions that we see in the world today.

Groups of people who are geographically and reproductively isolated from one another, tend to become different for a number of reasons, all of which may interact with each other. The environment itself may have a direct effect, for example, a diet deficient in iodine leads to high frequencies of goitres, and sunlight makes the skin darker. Environment factors may lead to different patterns of living, adapted to differences in the climate for example, and these in turn may lead to cultural differences. Differences in the ways of living and in language can also, however, arise simply by chance and historical accident. Cultures change continuously at varying rates, and it is not necessary to suppose that all the changes are adaptations to the prevailing environment or way of life. Exactly the same is true of inherited or genetic differences. In any population there are, as we shall see later, very many genetic differences between individuals. People may differ genetically, in such outwardly obvious features as their hair colour or their eye colour, or in unseen 'constitutional' characters, such as their blood types. Geneticists call the makeup of an individual his genotype. Thus people with blue eyes have a different genotype from those with brown or green eyes, at least as far as those genes determining eye colour are concerned. The frequencies of the various genotypes may be quite different in different populations. Thus the frequency of the blue-eyed genotype is, for example, much higher in Caucasian populations of European origin than in any other type of population. Genotype frequencies may change simply by chance, just like cultural differences, but such chance variations tend to be more important for small than for large populations.

Many, if not most, genetic changes are adaptive. By adaptive I mean that some one or more of those genotypes which are better suited to the environment leave more offspring or survive better, and so contribute relatively more to the next generation of individuals. This is the process of natural selection. Natural selection is the major agent of evolution because it is a major cause of changes in the genetic constitution of a population.

There are two major differences between cultural and inherited characteristics of populations. First, cultural characteristics tend to apply to all individuals of a population, whereas inherited differences are mostly measured by the frequency with which they occur in the population. Thus, while all members of a population will, for example, speak the same language and wear similar types of clothing, an inherited variation, such as blue versus green or brown eyes, may be found in a number of different populations, but it may occur with different frequencies in each of them. This means that a population is mostly characterized, not by being altogether of one or another genotype, but by the frequency with which the genotype is found in it. Blue eyes, for example, are certainly more common in northern than in southern Europe. But by no means all northerners are blue-eyed and some blue eyes will be found in the south. The second major difference between cultural and inherited characteristics is in the way that they are transmitted from individual to individual. Inherited traits are passed on from parents to offspring in accordance with Mendel's laws of genetics. Changes in the frequency of inherited traits depend on differences in the rate at which people of the various genotypes reproduce relative to one another. Genetic changes like these always take many generations, even when fairly strong natural selection is involved. Cultural characteristics, on the other hand, are not only passed from parents to offspring, but may be passed on from any one individual to another by word of mouth or by writing. So some cultural changes may be adopted quite quickly by a whole population. Transmission of culture is rather like transmission of an infection. Flu and cold epidemics spread very quickly, especially with the large amount of contact that people of all countries of the world now have with each other. In the same way cultural habits such as pop music preferences and clothing fashions may spread very quickly nowadays especially through the media of radio and television. However, other deep-rooted cultural characteristics of races and racial subgroups are much more difficult to change. These are the cultural patterns that are so resistant to alteration that they have the appearance of being innate; indeed, the difficulties in changing attitudes to school performance and in changing IQ in deprived populations, reflect in part, the difficulty in changing a cultural pattern.

Genetic polymorphism

Traditionally people have thought of human races in terms of outwardly obvious features such as skin colour, hair colour and texture, facial and other physical characteristics, whose inheritance cannot yet be explained in terms of simple gene differences whose pattern of occurrence in families follows Mendel's laws. Many common genetic differences are, however, known which are simply inherited but are not outwardly obvious and have no untoward effects. Most geneticists now would say that the only biologically valid approach to defining races is in terms of such simply inherited differences.

These simply inherited differences are mostly identified by laboratory tests on blood cells or serum. Perhaps the best known example of such differences are the ABO blood types. Blood donors always have to have their blood ABO typed for transfusion, so as to match their potential recipients. There are four common ABO types, A, B, AB and O which are genetically determined. An individual's type is determined by which versions of the gene responsible for making A, B or O substances he carries. Thus one form of the gene makes A, another B, and the third form neither. For example, individuals both of whose ABO genes are of the third form, are type O, while those with one A and one B gene, are type AB. (Remember that genes occur in pairs, one from each parent.) Geneticists call different versions of a gene, like those for the A or B substances, alleles.

Apart from identical twins, all people look different. The outward physical features by which we distinguish people are paralleled by simple inherited differences such as the ABO blood types. Such genetic traits are called polymorphisms, when the alternative versions of the genes that determine them, or the alleles as the geneticist calls them, each occur within a population with a substantial frequency. In the case of the ABO blood types, the allele A occurs in Caucasian populations with a frequency of 28 per cent, the B allele in 6 per cent and the O allele in 66 per cent. Over thirty such polymorphic genetic systems, including the blood groups, in which alternative genetic forms of a trait occur are already known and new ones are being discovered all the time. These thirty polymorphisms alone are enough to identify almost everyone as unique.

What proportion of genes occurring with a substantial frequency are polymorphic? The answer seems to be at least 30 per cent and could be even more.

So, the thirty or so polymorphisms we now know are a minute fraction of the total of more than 300,000 that must exist, taking into account the total number of genes in the human genome. The potential for genetic differences between individuals is truly staggering. The numbers of genetically different types of sperm or eggs which any one single individual could in principle produce, is many millionfold more than the number of humans that have ever lived. The polymorphisms so far discovered concern mainly chemical substances in the blood. Among those still to be discovered must surely be many which affect the chemical substances of the brain and involve behavioural differences. It is clear that the extraordinary genetic uniqueness of human individuals applies not just to the blood but to all physical, physiological and mental attributes. This enormous genetic variety within a population seems to be a property of almost all species.

Variation in gene frequencies between races

If we look at the blood types and other polymorphisms in races we find that the frequencies of polymorphic genes vary widely. Geneticists use these frequencies to define races. In Oriental populations, for instance, the frequency of the gene for the blood type B is 17 per cent while it is only 6 per cent in Caucasians (see figure 1). This means that type B individuals are generally three times as common in most of Asia than they are in Europe. All the known polymorphisms differ at least to some extent in their frequencies between different populations, though some are much more variable than others. One of the major tools in helping to work out the relationships between populations or races has been the analysis of the frequencies of genetic polymorphisms. In simple terms, the more similar are

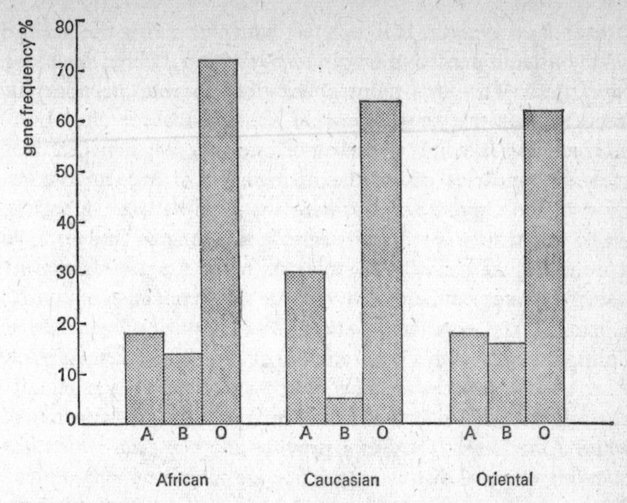

Figure 1 Frequencies of polymorphic genes among Africans, Caucasians and Orientals provide a means of differentiating these three races biologically. The columns represent the frequencies with which the genes of the A B O blood group system occur in these races

the polymorphic frequencies, the more closely related are the populations. A comparison of polymorphic frequencies in different races shows three very important features of the nature of genetic variation within and among them. Perhaps most important, the extent of variation *within* any population is usually far greater than the average difference *between* populations; in other words there is a great deal of overlap (for I Q see, for example, figure 5, page 104). Then, differences between populations and races are mostly measured by differences in the frequencies of the various genetic polymorphisms. They are not measured by whether or not a given gene is present. Any particular genetic combination may be found in almost any race, but the *frequency* with which it is found will vary from one race to another. Some genetic differences may simply be present or absent and so are, to some extent, characteristic of a race. Genes like this are clearly very useful in helping to delineate the human races, but they are very much the exception rather than the rule. Most genetic differences between populations can only be measured on average. It is this fact which underlies the need to distinguish differences among *individuals* from differences among *populations*. The final point about the study of polymorphic frequencies is that the distinction between races is often quite blurred. This is mainly the result of interbreeding between races at their boundaries, and of the mixing effect of large migrations. Even in the United States, where marriage between American blacks and whites is still quite rare, it is estimated that up to 30 per cent of the genes of the average black American from the northern states can be traced to white ancestry. This, of course, represents the cumulative effect of a number of generations during each of which a small amount of interbreeding took place. The American blacks are now clearly a new population genetically, which has been formed from a mixture of black Africans and white Americans. The Jews provide another example of the blending of racial distinctions. They are all presumably derived from one population, or at most a few closely related populations, of biblical times. Gene frequency studies, however, show that now Jewish populations in different parts of the world tend to resemble their surrounding populations at least as much, if not more, than they do each other. Such genetic studies, when combined with historical information, and sociological ideas of race and culture, can be very useful in understanding the origin of modern populations. Still remembering that most of the psychological and sociological studies of race are based on cultural determination of the boundaries, the definition of race in terms of differences in the frequencies of genetic polymorphisms is fairly arbitrary. How much difference does there have to be between populations before we call them different races? After all, even the people of, say, Lancashire and Yorkshire are likely to differ significantly in the frequency of at least some polymorphisms, but we should hardly refer to them as different races. On the other hand, most people would agree that the differences between the indigenous peoples of the major continents, such as the differences between Africans, Orientals and Caucasians, are obvious enough to merit the label race. Between these two extremes, however, lie a multitude of possibilities and it is largely a matter of taste as to whether one is a splitter or a lumper of population groups into races.

Inheritance of complex or quantitative characters

Heredity refers to the transmission of characteristics from parent to offspring. The primary biological functional unit of heredity is the gene, and the human genome – the complete set of genes which characterizes the biological inheritance of an individual – may consist of as many as five to ten million genes. Some of these genes can be individually identified by their patterns of inheritance in families, and the expression of many of these analysed at the biochemical level. The inheritance of differences between individuals which are known to be determined by one or a few genes can thus be reliably predicted and, in some cases, the biochemical or physiological basis for the inherited differences clearly established. Intelligence is, however, a composite and complex character, the expression of which must be dependant on a combination of the effects of environmental factors and the products of many different genes, each gene probably only having a small effect on measured IQ. The tools for dealing with the inheritance of such complex characters are necessarily complicated and still relatively ineffective.

As we have noted, intelligence must be a complex characteristic under the control of many genes. However, extreme deviations from normal levels, as in the cases of severe mental retardation, can sometimes be attributed to single gene differences. Such deviations can serve to illustrate important ways in which genetic factors can affect behaviour. Consider the disease phenylketonuria (PKU). Individuals with this receive from both of their parents a mutated version of the gene controlling the enzyme that converts one amino acid, phenylalanine, into another, tyrosine. The mutated gene allows phenylalanine to accumulate in the blood and in the brain, causing mental retardation. The accumulation can, to some extent, be checked early in life by a diet deficient in phenylalanine.

The difference between the amounts of phenylalanine in the blood of people with PKU and that in the blood of normal people, which is closely related to the primary activity of the gene causing PKU, clearly creates two genetic classes of individuals. When such differences are compared with differences in IQ, there is a slight overlap, but individuals afflicted with PKU can be distinguished clearly from normal individuals. It is,

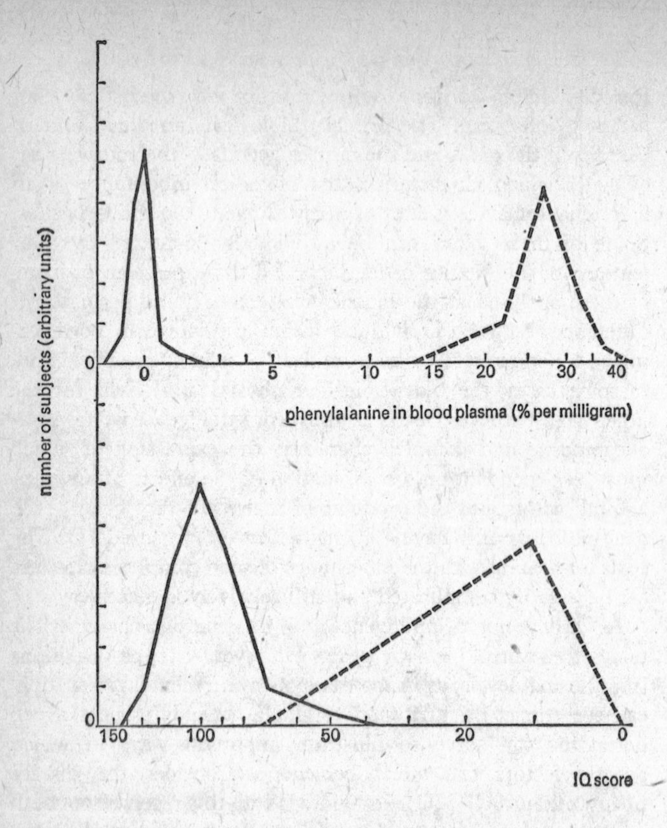

Figure 2a Phenylalanine levels in blood plasma shown in first set of curves (top) distinguish those who carry a double dose of the defective gene that causes high phenylalanine levels (broken curve), a condition called phenylketonuria, from those with normal phenylalanine levels (solid curve). Second set of curves (bottom) shows that this genotype has a direct effect on intelligence: phenylketonurics (broken curve) have low I Qs because accumulation of phenylalanine and its by-products in blood and nerve tissue damages the brain. Individuals with functioning gene (solid curve) have normal I Qs.

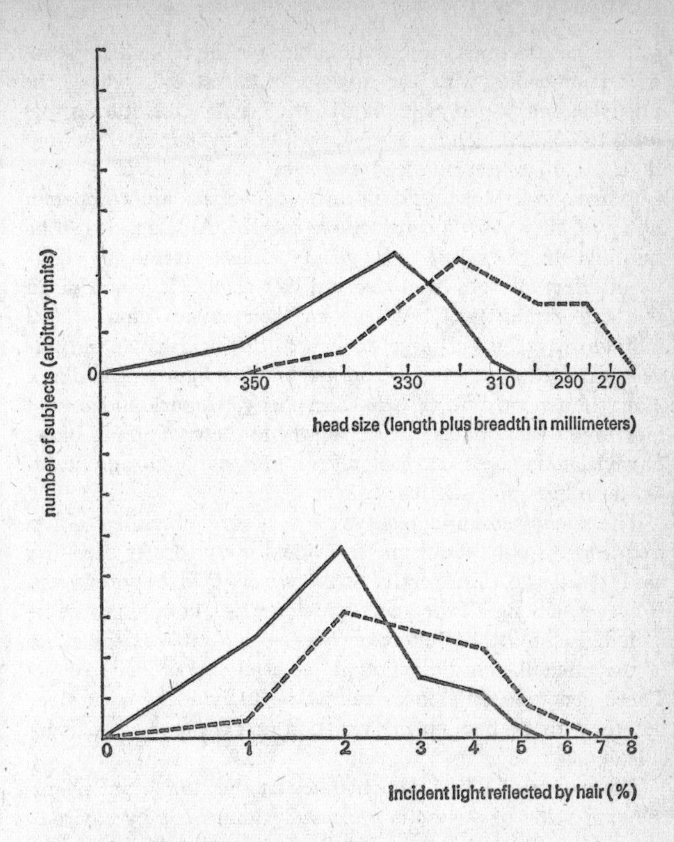

Figure 2b In the third set of curves (top) phenylalanine levels are related to head size (displayed as the sum of head length and breadth), and in the fourth set (bottom) phenylalanine levels are related to hair colour (displayed as the percentage of light with a wavelength of 700 millimicrons reflected by the hair). In both cases it is obvious that the phenylketonuric genotype has a significant effect on each of these characteristics: the reflectance is greater and the head size is smaller (broken curves) among phenylketonurics than they are among normal individuals (solid curves). Yet the distribution of these characteristics is such that they cannot be used to distinguish those afflicted with phenylketonuria from those who are not.

indeed, routine today to test all babies for the tell-tale presence of ketane bodies with a nappy or urine test that reveals the phenylketonuric genotype, that is, the genetic constitution that leads to PKU which is associated with extreme mental retardation. If differences in head size and hair colour in phenylketonuric individuals and normal individuals are compared, however, they show a considerable overlap. Although it can be said that the phenylketonuric genotype has a statistically significant effect on both head size and hair colour, it is not, given the large variations of head size and hair colour among PKU individuals, an effect large enough to distinguish the phenylketonuric genotype from the normal one (see figures 2a and 2b). Thus the genetic difference between phenylketonuric and normal individuals contributes in a major way to the variation in blood phenylalanine levels but has only a minor, although significant, effect on head size and hair colour.

The phenylketonuric genotype is very rare, occurring with a frequency of only about one individual in 10,000. It therefore has little effect on the overall distribution of IQ in the population. However, among all the genes which are polymorphic must be included many whose effect on IQ is comparable to the effect of the phenylketonuric genotype on head size or hair colour. These genotype differences cannot be individually identified, but their total effect on the variation of IQ may be considerable.

The nature of PKU demonstrates another important point: the expression of a gene is profoundly influenced by environment. Phenylketonuric individuals show appreciable variation. This indicates that the genetic difference involved in PKU is by no means the only factor, or even the major factor, affecting the level of phenylalanine in the blood. It is obvious that dietary differences have a large effect, since a phenylalanine-deficient diet brings the level of this amino acid in the blood of a phenylketonuric individual almost down to normal. If an individual receives the phenylketonuric gene from only one parent, his mental development is not likely to be clinically affected. Nevertheless, he will tend to have higher than normal levels of phenylalanine in his blood. The overall variation in phenylalanine level is therefore the result of a combination of genetic factors and environmental factors. Measuring the relative contribution of genetic factors to the overall variation is thus equivalent to measuring the relative importance of genetic differences in determining this type of quantitative variation.

Characters determined by the joint action of many genes, like height and IQ, are often called quantitative characters because they are measured on a continuous scale. They are much more susceptible to environmental influences than are polymorphisms such as the blood groups which seem to be pretty well independent of all environmental factors. Because the contribution of individual genes to such characters cannot easily be recognized, one has to resort to complex statistical analyses to sort out the relative contributions of heredity and environment. These analyses try to assess the extent to which relatives tend to be more like each other than they are to unrelated or to more distantly related people.

The occurrence of identical and non-identical twins is a sort of natural experiment that illustrates very simply the way in which environmental and genetic factors can be separated. Identical (or monozygotic = MZ) twins are derived from a single fertilized egg and so are identical genetically. Any differences between them must, therefore, be due to the environment. Non-identical (or dizygotic = DZ) twins come from two eggs fertilized at the same time by different sperm and so are as different, or alike, genetically as any brothers or sisters. Both types of twins are usually brought up within a family and so, on average, are subject to more or less the same types of environmental variation. The extent to which non-identical twins differ more than identical twins is thus an indication of the importance of genetic factors in differentiating the non-identical twins. Differences in IQ, among non-identical pairs show a greater spread than those among identical pairs, indicating that the genetic diversity among the non-identical pairs adds to the purely environmental differences which differentiate the identical pairs of twins. A comparison of the average difference between members of individual identical pairs and the average difference between members of non-identical pairs, could in principle be taken as a measure of the relative importance of genetic and environmental factors. (For statistical reasons, it is usually better to consider not the mean differences but their squares. These are directly related to the *variance*, which is a well-known statistical measure of the spread of a distribution.)

There are two major contrasting reasons why such a simple measure is not entirely satisfactory. First, the difference between members of a dizygous pair represents only a fraction of the genetic differences that can exist between two individuals. Dizygous twins are related to each other as two siblings are; therefore they are more closely related than two individuals taken at random from a population. This implies a substantial reduction (roughly by a factor of two) in the average genetic difference between dizygous twins compared with that between two randomly chosen individuals. Secondly the environmental difference between members of a pair of twins encompasses only a fraction of the total environmental difference that can exist between two individuals, namely the difference between individuals belonging to the same family. This does not take into account differences among families, which are likely to be large. Within the family the environmental differences between twins are limited. For instance, the effect of birth order is not taken into account. Differences between ordinary siblings might therefore tend to be slightly greater than those between dizygous twins. It also seems possible that the environmental differences between monozygous twins, who tend to establish special relations with each other, are not exactly comparable to those between dizygous twins. In short, whereas the contrast between monozygous and dizygous twin pairs minimizes genetic differences, it also tends to minimize environmental differences.

In order to take account of such difficulties one must try to use all available comparisons between relatives of various types and degrees, of which twin data are only a selected case. For technical reasons one often measures similarities rather than differences between two sets of values such as parent IQs and offspring IQs. Such a measure of similarity is called the correlation coefficient. It is equal to 1 when the pairs of values in the two sets are identical or, more generally, when one value is expressible as a linear function of the other. The correlation coefficient is 0 when the pairs of measurements are completely independent, and it is intermediate if there is a relation between the two sets such that one tends to increase when the other increases.

The mean observed values of the correlation coefficient between parent and child IQs and between the IQs of pairs of siblings, are close to 0·5. This is the value one would expect on the basis of the simplest genetic model in which the effects of any number of genes determine IQ and there are no environmental influences or complications of any kind. It seems probable, however, that the observed correlation of 0·5 is coincidental. Complicating factors such as different modes of gene action, tendencies for like to mate with like and environmental correlations among members of the same family must just happen to balance one another almost exactly to give a result that agrees with the simplest theoretical expectation. If we ignored these complications, we might conclude naively (and in contradiction to other evidence, such as the observations on twins) that biological inheritance of the simplest kind entirely determines IQ.

Heritability of IQ

We need a means of determining the relative importance of environmental factors and genetic factors, taking account of several of the complications. In theory this measurement can be made by computing the quotients known as heritability estimates. To understand what such quotients are intended to measure, consider a simplified situation. Imagine that the genotype of each individual with respect to genes affecting IQ can be identified. Individuals with the same genotype can then be grouped together. The differences among them would be the result of environmental factors, and the spread of the distribution of such differences could then be measured. Assume for the sake of simplicity that the spread of IQ due to environmental differences is the same for each genotype. If we take the IQs of all the individuals in the population, we obtain a distribution that yields the total variation of IQ. The variation within each genotype is the environmental component. The difference between the total variation and the environmental component of variation leaves a component of the total variation that must be accounted for by genetic differences. This component, when expressed as a fraction of the total variance, is one possible measure of heritability.

In practice, however, the estimation of the component of the total variation that can be accounted for by genetic differences (from data on correlations between relatives) always depends on the construction of specific genetic models, and is therefore subject to the limitations of the models. One problem lies in the fact that there are a number of alternative definitions of heritability depending on the genetic model chosen, because the genetic variation may have many components that can have quite different meanings. A definition that includes only those parts of the genetic variation generally considered to be most relevant in animal and plant breeding is often used. This is called heritability in the narrow sense. If all genetic sources of variation are included, then the heritability estimate increases and is referred to as heritability in the broad sense.

The differences between these estimates of heritability can be defined quite precisely in terms of specific genetic models. The resulting estimates of heritability, however, can vary considerably. Typical heritability estimates for I Q (derived from the London population in the early 1950s, with data obtained by Sir Cyril Burt) give values of 45 to 60 per cent for heritability in the narrow sense and 80 to 85 per cent for heritability in the broad sense.

A further major complication for such heritability estimates has the technical name 'genotype–environment interaction'. The difficulty is that the realized I Q of given genotypes in different environments cannot be predicted in a simple way. A given genotype may develop better in one environment than in another. In man there is no way of controlling the environment. Even if all environmental influences relevant to behavioural development were known, their statistical control by appropriate measurements and subsequent statistical analysis of the data would still be extremely difficult. It should therefore be emphasized that, because estimates of heritability depend on the extent of environmental and genetic variation that prevails in the population examined at the time of analysis, they are not valid for other populations or for the same population at a different time.

The investigation of the same genotype or similar genotypes in different environments can provide valuable controls over environmental effects. In man this can be done only through the study of adopted children. A particularly interesting type of 'adoption' is that in which monozygous twins are separated and reared in different families from birth or soon afterwards. The outcome is in general a relatively minor average decrease in similarity. Following the same line of reasoning, the similarity between foster parents and adopted children can be measured and contrasted with that between biological parents and their children (see figure 3). The results show that, though the correlation between foster parents and their children is significantly greater than 0, it is undoubtedly less than that between biological parents and their offspring. However, a complete analysis of such data is difficult because children are not adopted at random and so even adoption does not provide a convincing control over the environment. Nevertheless, on the basis of all the available data and allowing for the limitations to its interpretations, the heritability of IQ is still fairly high. It must be emphasized however, that the environmental effects in essentially all the studies done so far are limited to the differences among and within families of fairly homogenous sections of the British or United States populations. The results cannot therefore be extrapolated to the prediction of the effects of greater differences in environment or to other types of differences.

IQ and social class

There are significant differences in mean IQ among the various social classes. One of the most comprehensive and widely quoted studies of such differences and the reasons for their apparent stability over the years was published by Burt in 1961 (see figure 4). His data come from schoolchildren and their parents in a typical London borough. Socio-economic level was classified, on the basis of type of occupation, into six classes. These range from Class 1, including 'university teachers, those of similar standing in law, medicine, education or the Church and the top people in commerce, industry or civil service', to Class 6, including 'unskilled labourers and those employed in coarse manual work'. There are four main features of these data:

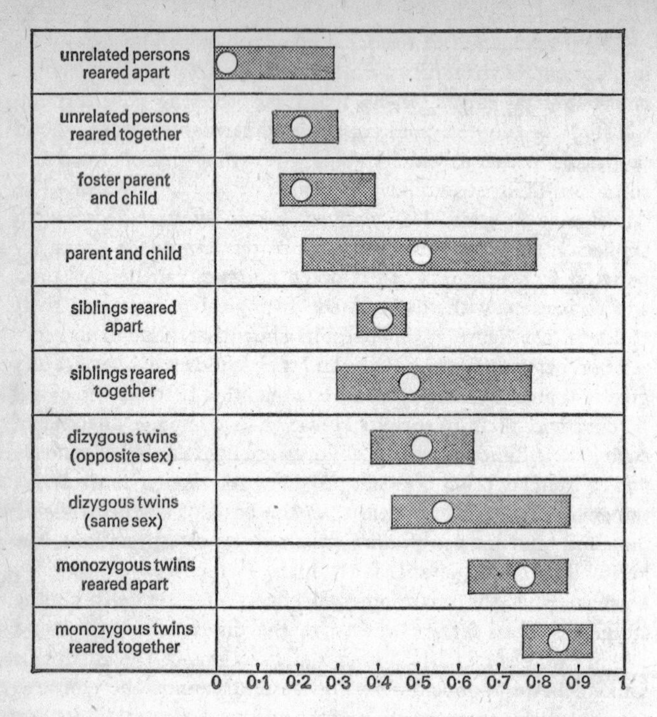

Figure 3 Correlation coefficients are representative of similarities and differences. A coefficient of 1 indicates that the two classes compared are identical. 0 indicates independence of one value from the other. The horizontal bars represent the range of differences in coefficients found in the relationships shown. The mean of each range is represented by the circle in each bar. A mean coefficient of 0·50 is that which would be expected if there were no environmental effects in IQ. As other evidence indicates that environment exerts a significant effect, these calculations must be further refined. These data are extracted from the work of Erlenmeyer-Kimling and Jarvik.

1. Parental mean IQ and occupational class are closely related. The mean difference between the highest and the lowest class is over 50. Although occupational class is determined mostly by the father, the relatively high correlation between the IQs of

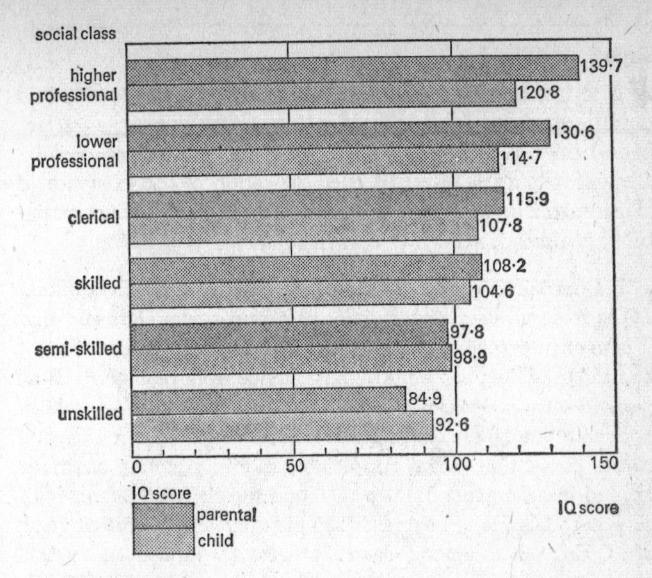

Figure 4 An English study by Burt indicates that intelligence and social class are closely related. The darker bars represent parental I Qs for the social classes while the lighter bars indicate the I Qs of their children. The phenomenon of children of above average parents having lower I Q scores is known as regression to the mean, and children of low-I Q parents exhibit an upward trend towards the population mean.

husband and wife (about 0·4) contributes to the differentiation among the classes with respect to IQ.

2. In spite of the significant variation between the parental mean IQs, the residual variation in IQ among parents within each class is still remarkably large. The mean standard deviation of the parental IQs for the different classes is 8·6, almost three-fifths of the standard deviation for the entire group. That standard deviation is contrived in test construction to be about 15.

3. The mean IQ of the offspring for each class lies almost exactly between the parental mean IQs and the overall population mean IQ of 100. This is expected because it is only another way of looking at the correlation for IQ between parent and

child, which as we have already seen tends to be about 0·5 in any given population.[1]

4. The last important feature of the data is that the standard deviations of the IQ of the offspring, which average 13·2, are almost the same as the standard deviation of the general population, namely 15. This is another indication of the existence of considerable variability of IQ within social classes. Such variability is almost as much as that in the entire population.

The most straightforward interpretation of these data is that IQ is itself a major determinant of occupational class and that it is to an appreciable extent inherited (although the data cannot be used to distinguish cultural inheritance from biological). Burt pointed out that, because of the wide distribution of IQ within each class among the offspring and the regression of the offspring to the population mean, appreciable mobility among classes is needed in each generation to maintain the class differences with respect to IQ. He estimated that to maintain a stable distribution of IQ differences among classes, at least 22 per cent of the offspring would have to change class, mainly as a function of IQ, in each generation. This figure is well below the observed intergenerational social mobility in Britain, which is about 30 per cent.

Fears that there may be a gradual decline in IQ because of

1. We must note an important fallacy in Eysenck's (and others') argument that 'regression presents strong evidence for genetic determination of IQ differences'. In any multi-causal system, such as that involved in the determination of IQ, there will be many factors which will tend to increase performance, and others that work to decrease it. In the average situation these will balance and give the mean IQ of 100. Less commonly they will not balance and we will get extreme IQs, well above or below the mean. These extreme cases arise from relatively rare interactions of genes and environment where (at least statistically) most of the causal factors are pushing in one direction. It is extremely unlikely that this situation will hold for children of very high or very low IQ parents. Things will tend to be more equal and causal factors more nearly balance, so the children's IQ will regress to the mean and the largest regression will be seen in the children of extreme IQ parents. All the presence of regression tells us is that the system is multi-causal. It says nothing about the origin of these causes. Indeed, it is theoretically possible to have regression to the mean in an entirely environmentally controlled system. [Eds.]

an apparent negative correlation between IQ and fertility have been expressed ever since Francis Galton pointed out this correlation for the British ruling class in the second half of the nineteenth century. If there were such a persistent association, if IQ were at least in part genetically determined and if there were no counteracting environmental effects, such a decline in IQ could be expected. The fact is that no significant decline has been detected so far. The existing data, although they are admittedly limited, do not support the idea of a persistent negative correlation between IQ and overall reproductivity.

IQ–Race differences

The average frequency of marriages between blacks and whites throughout the US is still only about 2 per cent of the frequency that would be expected if marriages occurred at random with respect to race. This reflects the persistent high level of reproductive isolation between the races, in spite of the movement in recent years towards a strong legal stance in favour of desegregation. Hawaii is a notable exception to this separation of the races, although even there the observed frequency of mixed marriages is still only 45 to 50 per cent of what would be expected if matings occurred at random.

Many studies have shown the existence of substantial differences in the distribution of IQ in US blacks and whites. Such data were obtained in an extensive study published by Wallace A. Kennedy of Florida State University and his co-workers in 1963, based on IQ tests given to 1800 black children in elementary school in five south-eastern states (Florida, Georgia, Alabama, Tennessee and South Carolina) (see figure 5). When the distribution these workers found is compared with a 1960 sample of the US white population, striking differences emerge. The mean difference in IQ between blacks and whites is 21·1, whereas the standard deviation of the distribution among blacks is some 25 per cent less than that of the distribution among whites (12·4 *v.* 16·4). As one would expect there is considerable overlap between the two distributions, because the variability for IQ within any population is (like the variability for most characteristics) substantially greater than the variability between any two populations. Nevertheless,

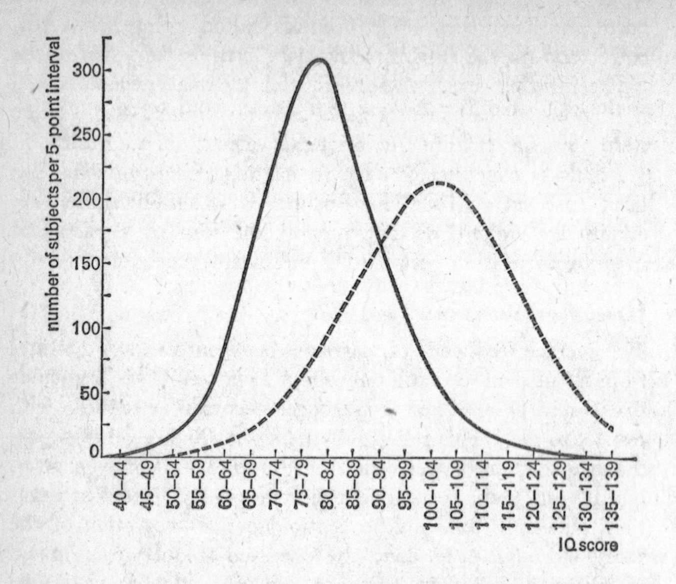

Figure 5 IQ difference between blacks and whites in the USA emerges from a comparison of the IQ distribution in a representative sample of whites (broken curve) with the IQ distribution among 1800 black children in the schools of Alabama, Florida, Georgia, Tennessee and South Carolina (black curve). Wallace A. Kennedy of Florida State University, who surveyed the students' IQ, found that the mean IQ of this group was 80·7. The mean IQ of the white sample is 101·3, a difference of 21·1 points. The two samples overlap distinctly but there is also a sizable difference between the two means. Other studies show a difference of 10 to 20 points, making Kennedy's result one of the most extreme reported.

95·5 per cent of the blacks have an IQ below the white mean of 101·8 and 18·4 per cent have an IQ of less than 70. Only 2 per cent of the whites have IQs in the latter range.

Reported differences between the mean IQs of blacks and whites generally lie between 10 and 20, so that the value found by Kennedy and his colleagues is one of the most extreme reported. The difference is usually less for blacks from the northern states than it is for those from the southern states, and clearly it depends heavily on the particular populations tested. One well-known study of army 'alpha' intelligence-test results, for example, showed that blacks from some northern states achieved higher average scores than whites from some southern states, although whites always scored higher than blacks from the same state. There are many uncertainties and variables that influence the outcome of I Q tests, but the observed mean differences between U S blacks and whites are undoubtedly more or less reproducible and are quite striking.

There are two main features that clearly distinguish IQ differences among social classes described above from those between blacks and whites. First, the IQ differences among social classes relate to the environmental variation within the relatively homogeneous British population. It cannot be assumed that this range of environmental variation is comparable with the average environmental difference between black and white Americans. Secondly, and more important, these differences are maintained by the mobility among occupational classes that is based to a significant extent on selection for higher IQ in the higher occupational classes. There is clearly no counterpart of this mobility with respect to the differences between US blacks and whites; skin colour effectively bars mobility between the races.

The arguments for a substantial genetic component in the IQ difference between the races assume that existing heritability estimates for IQ can reasonably be applied to the racial difference. These estimates, however, are based on observations within the white population. We have emphasized that heritability estimates apply only to the population studied and to its particular environment. Thus the extrapolation of existing heritability estimates to the racial differences assumes that the environmental differences between the races are comparable to the environmental variation within them. Let us consider a simple model example which shows that there is no logical connection between heritability as determined within races and the genetic differences between them. Suppose we take a bag of seed collected from a field of wheat and sow one handful on barren, stony ground and another one on rich, fertile ground. The seeds sown on fertile ground will clearly grow more vigorously and give a much higher yield per plant than that sown on barren ground. If we were to study the extent to which individual differences in yield between the plants grown on the fertile ground were genetically determined, we should expect to find comparable genetic differences to those which existed between the plants grown in the original wheat field from which our bag of seed was collected. The same would be true for the differences between the individual plants growing on the barren ground. The fact that there are genetic differences between these plants both on the fertile and on the barren ground clearly does not have anything to do with the overall differences between the two sets of plants grown in these two very different environments. This we know must be due to the differences in the soil. The genetic stock from which all the plants were derived was, after all, the same. It was the original one bag of seed. The same logic must apply to the human situation. How can we be sure that the different environments of US blacks and whites are not comparable to the barren and fertile soils? Our original bag of human genes may not be as uniform as the wheat field, but why should the differences between the original human samples, the Africans and the Caucasians, have anything to do with conventional IQ measurements? Whether or not the variation in IQ within either race is entirely genetic or entirely environmental has no bearing on the question of the relative contribution of genetic factors and environmental factors to the differences between the races.

IQ and environment

A major argument given by Jensen, Eysenck and others in favour of a substantial genetic component to the IQ difference is that it persists even when comparisons are made between US blacks and whites of the same socio-economic status. This status is defined in terms of schooling, occupation and income, and so it is necessarily a measure of at least a part of the environmental variation, comparable to the class differences we have discussed here.

Taken on face value – that is, on the assumption that status is truly a measure of the total environment – these data would indicate that the IQ difference is genetically determined. It is difficult to see, however, how the status of blacks and whites can be compared. The very existence of a racial stratification correlated with a relative socio-economic deprivation makes this comparison suspect. Black schools are well known to be generally less adequate than white schools, so that equal numbers of years of schooling certainly do not mean equal educational attainment. Wide variation in the level of occupation must exist within each occupational class. Thus one would certainly expect, even for equivalent occupational classes, that the black level is on the average lower than the white. No amount of money can buy a black person's way into a privileged, upper-class, white community, or buy off more than two hundred years of accumulated racial prejudice on the part of the whites, or reconstitute the disrupted black family, in part culturally inherited from the days of slavery. It is impossible to accept the idea that matching for status provides an adequate, or even a substantial, control over the most important environmental differences between blacks and whites.

Let us consider just two examples of environmental effects on IQ which show both how complicated is the environment and how large can be the effects of environmental differences. The first concerns a comparison of the IQs of twins and triplets as compared to single births. A number of studies have shown that twins have an IQ that is systematically about 5 points lower than non-twins. This reduction seems to be independent of socio-economic variables and of such other factors as parental age, birth order, overall family size, gestation time and birth weight as has been shown particularly convincingly by a recent study carried out in Birmingham, England, by Record, McKeown and Edwards. These authors based their study on the results of verbal reasoning tests given in the 11+ examinations during the period from 1950 to 1954. The average scores for 48,913 single births and 2164 twin births were 100·1 and 95·7 respectively. Furthermore the average score for 33 triplets was 91·6, another 4 points lower, while the average score for 148 twins, whose co-twins were stillborn or died within four weeks after birth was 98·8. This, they point out, is only very little lower than the score of 99·5, which is obtained by standardizing the data to the maternal ages and birth ranks observed for these 148 twins. The environmental factor involved in this remarkable effect on IQ must be post-natal and may well have something to do with the reduced attention parents are able to give each of two very young children born at the same time. This one subtle factor in the familial environment, which clearly is not reflected in standard measurements of socio-economic status, has an effect on IQ which is about one-third of the overall average difference between US blacks and whites. Measuring the environment only by standard socio-economic parameters, is a little bit like trying to assess the character of an individual by his height, weight and eye colour.

The second example of a major environmental effect comes from a well known study of adopted children. Skodac and Skeels studied the IQs of a series of white children placed into adopted homes through Orphans' Home Institutions in Iowa, mostly before the infants were six months old, and compared these IQs with the IQs of their foster parents and of their biological parents. As in other similar studies they found a higher correlation between the IQs of the adopted children and their biological parents than with their foster parents. Most strikingly, however, they found that while the mean IQ of the 63 adopted children who had been followed through until they were about thirteen to fourteen years old was 106, the mean IQ of the biological mothers of these 63 children was only 85·5, a difference of fully 20 points! Even if one assumes that the biological mothers came from a low socio-economic group and that their husbands had the population average IQ of 100 and that IQ is completely genetically determined, the expected average IQ of the children would be only one half $(100+85{\cdot}5) = 92{\cdot}75$. The difference between expected and observed is still $106-92{\cdot}75 = 13{\cdot}25$ points, which is just about the same as the average US black-white IQ difference. The adoptive homes were strongly biased towards the upper socio-economic strata. This study thus shows what a striking effect an improved home background can have on IQ. In Skodak and Skeels's own words, 'the implications for placing agencies justify a policy of early placement in adoptive homes offering emotional warmth and security in an above average educational and social setting.'

Why should there be a genetic component to the race–IQ difference?

Jensen has stated that because the gene pools of whites and blacks are known to differ and 'these genetic differences are manifested in virtually every anatomical, physiological and biochemical comparison one can make between representative samples of identifiable racial groups . . . there is no reason to suppose that the brain should be exempt from this generalization.' But there is no *a priori* reason why genes affecting IQ, which differ in the gene pools of blacks and whites, should be such that on the average whites have significantly more genes increasing IQ than blacks do. On the contrary, one should expect, assuming no tendency for high IQ genes to accumulate by selection in one race or the other, that the more polymorphic genes there are that affect IQ and that differ in frequency in blacks and whites, the less likely it is that there is an average genetic difference in IQ between the races. The same argument applies to the differences between any two racial groups.

Since natural selection is the principal agent of genetic change, is it possible that this force has produced a significant IQ difference between American blacks and whites? Using the simple theory with which plant and animal breeders predict responses to artificial selection, one can make a rough guess at the amount of selection that would have been needed to result in a difference of about 15 IQ points, such as exists between blacks and whites. The calculation is based on three assumptions: that there was no initial difference in IQ between Africans and Caucasians, that the heritability of IQ in the narrow sense is about 50 per cent and that the divergence of black Americans from Africans started with slavery about two hundred years, or seven generations, ago. This implies a mean change in IQ of about 2 points per generation. The predictions of the theory are that this rate of change could be achieved by the complete elimination from reproduction of about 15 per cent of the most intelligent individuals in each generation. There is certainly no good basis for assuming such a level of selection against IQ during the period of slavery.

Eysenck has actually suggested as one basis for genetic differences that significant selection might have taken place during the procurement of slaves in Africa. He proposes, for example, that the more intelligent were less likely to be caught or that the less intelligent were the ones that were sold off by their tribal chiefs. If one makes the extreme assumption that this initial selection is the sole cause of a 15 point IQ difference between US blacks and whites, then following the same lines of analysis that were used above, the slaves that were caught or sold must have been in the bottom 5 per cent as far as IQ is concerned. This would seem to indicate that the tribal chiefs knew how to administer and interpret IQ tests almost as well as we do today! It is clearly very difficult, if not impossible, to assess the role that natural selection might have played in accentuating IQ differences between the races. Certainly any hypothesis one chooses to put forward is in the realm of unsubstantiated speculation and cannot be claimed as even suggestive evidence for a substantial genetic component to the race–IQ difference.

One approach to studying this question that has been suggested by Shockley, Eysenck and others is to correlate IQ measurements with an assessment of the proportion of Caucasian genes in different samples of US blacks. The proportion of Caucasian genes in a sample of US blacks can be estimated from a comparison of the frequencies of polymorphic genes in Caucasians, African blacks and US blacks. This approach would mean that groups with different degrees of admixture of 'white' and 'black' genes would have to be studied for their IQ. It is known, however, that the degree of admixture varies from less than 10 per cent in some southern states to more than 30 per cent in some northern states. For this knowledge to be useful, one has to assume that the environment is the same for all these populations and that there is no correlation between differences in the environment and the extent of the black–white admixture. One example of data that shows quite clearly that this is not true has been analysed by Spuhler and Lindzey. They pointed out that for both blacks and whites, the mean values per state of US army alpha intelligence-test scores obtained during the First World War correlated precisely with the *per capita* state expenditure, at that time, on education. The states with a mean expenditure of less than $5 per head (Arkansas, Florida, Georgia, Kentucky, Louisiana, Mississippi, North and South Carolina, Tennessee, Texas and Virginia – *all* southern states) gave overall mean alpha scores of 50·2 for whites and 24·9 for blacks. The states with a mean expenditure of more than $10 (Illinois, Indiana, Kansas, New Jersey, New York, Ohio and Pennsylvania) gave mean alpha scores of 65·3 for whites and 44·9 for blacks. Southern whites hardly scored better than northern blacks, as pointed out long ago by J. B. S. Haldane. Any hope of using such data for genetic studies is clearly out of the question.

The only approach applicable to the study of the IQ difference between the races is that of working with black children adopted into white homes and vice versa. The adoptions would, of course, have to be at an early age to be sure of taking into account any possible effects of the early home environment. The IQs of black children adopted into white homes would also have to be compared with those of white children adopted into comparable white homes. To our knowledge no scientifically adequate studies of this nature have ever been undertaken. It is questionable whether or not such studies could be done in a reasonably controlled way at the present time. Even if they could, they would not remove the effects of prejudice directed against black people in most white communities. It therefore seems that the question of a possible genetic basis for the race–IQ difference will be almost impossible to answer satisfactorily before the environmental differences between US blacks and whites have been substantially reduced.

Jensen has stated on the basis of his assessment of the data (and Eysenck quotes him):

> so all we are left with are various lines of evidence no one of which is definitive alone, but which, viewed all together, make it a not unreasonable hypothesis that genetic factors are strongly implicated in the average Negro–white intelligence difference. The preponderance of evidence is, in my opinion, less consistent with a strictly environ-

mental hypothesis than with a genetic hypothesis, which of course does not exclude the influence of environment or its interaction with genetic factors.

My assessment of the evidence made together with my colleague Professor Cavalli-Sforza and published in an article in the *Scientific American* – and this is an assessment that we share with many of our geneticist colleagues – is that we simply do not have enough evidence at present to resolve the question. It seems to us that differences in I Q, for instance, between American blacks and whites, could be explained by environmental factors, many of which we still know nothing about. This does not mean that we exclude the possibility that there might be a genetic component to such a mean I Q difference. We simply maintain that currently available data are inadequate to resolve this question in either direction and that we cannot see how the question could be satisfactorily answered using presently available techniques.

What use can be made of knowledge concerning genetic components to race–I Q differences?

The Nobel prize-winning U S physicist, William Shockley has repeatedly asked for a major expenditure of funds directed specifically at finding the answer to the question of a genetic component to the race I Q difference and other similar questions, because of their practical importance. He has in fact said in this context:

> I believe that a nation that achieved its ten-year objective of putting a man on the moon can wisely and humanely solve its human quality problems once the objective is stated and relevant facts courageously sought.

No one surely should argue against the need for a better scientific understanding of the basis of intellectual ability and the benefits to society that might accrue from such an understanding. But why concentrate this effort on the genetic basis for the race–I Q difference? Apart from the intrinsic, almost insurmountable difficulties in answering this question at the present time, it is not in any way clear what practical use could be made of the answers. Perhaps the only practical argument is that, since the question that the difference is genetic has been raised, an attempt should be made to answer it. Otherwise those who now believe that the difference is genetic will be left to continue their campaigns for an adjustment of our educational and economic systems to take account of 'innate' racial differences.

A demonstration that the difference is not primarily genetic could counter such campaigns. On the other hand, an answer in the opposite direction should not, in a genuinely democratic society free of race prejudice, make any difference. Our society professes to believe there should be no discrimination against an individual on the basis of race, religion or other *a priori* categorizations, including sex. Our accepted ethic holds that each individual should be given equal and maximum opportunity, according to his or her needs, to develop to his or her fullest potential. Surely innate differences in ability and other individual variations should be taken into account by our educational system. These differences must, however, be judged on the basis of the individual and not on the basis of race. To maintain otherwise indicates an inability to distinguish differences among individuals from differences among populations.

Further reading

L. L. Cavalli-Sforza and W. F. Bodmer, *The Genetics of Human Population*, Freeman, 1971.

T. Dobzhansky, *Mankind Evolving*, Yale University Press, 1962.

I. M. Lerner, *Heredity, Evolution and Society*, Freeman, 1968. This book is particularly suitable for the non-biologist.

J. Maynard-Smith, *The Theory of Evolution*, Penguin, 1966.
Myths in Human Biology, BBC Publications, 1972. A collection of transcripts from recent radio talks.

6 Diversity: A Developmental Perspective

John Hambley

John Hambley took a degree in biological sciences at the University of Adelaide and did teaching and research in genetics in America before joining The Open University where he now researches into the effects of early experiences on brain chemistry.

Introduction

It is generally accepted that Charles Darwin initiated a revolution in the way we think about nature and ourselves. The idea of change, the appreciation of chance as a factor in generating order, the rejection of extreme typological approaches to the sample of nature we are dealing with at any one time – all these are now part of the intellectual framework of every biologist. One of our responsibilities as biologists is to facilitate the diffusion and appreciation of these subtleties to others; for we live in what we might call a typological world. This is perhaps a reflection of the necessity, in written language and discussion, of adopting various classificatory schemes for dealing with the variability around us; but we should always be careful to see these as conveniences for the representation of reality. The question is, can we reach a scientific understanding of variability? This, I believe, is the importance of the biological perspective; it is a framework for assessing the significance of variability, of uniqueness. Even so, biologists talk of the unity represented by this immense diversity. We obviously need to examine some of the implications of a biological approach to mankind and how these bear upon the matters under discussion in this book.

But here perhaps one should go on record as saying that to discuss IQ is not to agree that the rich theoretical construct of human intelligence can be adequately represented by a score derived from simple selective problem-solving. Such matters are, however, discussed elsewhere.

In this chapter I propose some biological views which, it is hoped, will give the reader an insight into the significance of individual differences. It will be necessary to introduce some new terms but I feel this excursion into semantics will clear up some confusions and enable the reader to better appraise what others have said. No apology is offered for including biological considerations here. It is not an irrelevant intrusion, for the whole debate has often been marred by misleading platitudes about nature and nurture and the railings of extreme empiricists and crude nativists against one another.

The basis of diversity

Genes, you will remember, are the basic particles of heredity and are located on the chromosomes found in the nucleus of each cell. The chromosomes themselves are differentiated longitudinally and individual genes are arranged in a linear sequence down the chromosome. Furthermore, each gene in a cell occurs as one of a pair; there are thus two sets of genes, one derived from each parent. Each gene or allele in any one pair can be the same (as in the homozygous condition) or there may be alternative forms of the one gene (the heterozygous condition). Because of the large number of genes, their multiple alternate forms, the shuffling of these units in producing the ensemble of genes in each egg and sperm and the chance uniting of just two such gene combinations at conception, there is tremendous scope for variation. In fact we can say that with the exception of identical twins each individual is genetically unique. This unique set of genes which an individual receives we call his genotype, and this is fixed at conception. From any particular genotype an individual organism will develop; phenotype is the term used to describe the sum of the characteristics manifested in the individual. As such it is not something very distinct or fixed. It will vary with time and must deal separately with various aspects of the organism since it is not possible to have a single total description. But we must look more closely at this process of development, for it results in an essential indeterminacy in the relation between phenotype and genotype.

The proximate action of each gene is to produce a protein: but not all genes, perhaps very few, are active at any one time. From the moment of conception genes are being switched on and off; some function only for a very short time, others perhaps for life. It is the integration of this switching on and off and the interaction of all the products in forming new cells and structures that constitutes development. Another term which needs to be introduced here is environment. As the developmental biologist conceives of it, environment is everything external to the genotype (fixed at conception) mediating its expression at any one time. This includes the products, immediate and remote, of other genes, as well as the external environment of the organism. Indeed, effects of the external environment may in part be mediated at the genic level. For example, it is known that hormones which can be produced in response to external stimuli may be involved in switching genes on and off. An important point to make here is that, although environment influences this complex elaboration, this does not imply Lamarckism, the biological inheritance of acquired characters. For at this time there is no evidence that environment can cause directed adaptive change in the genetic material itself; the mediation is by changes in the rates and numbers of genes acting over time.

Environment then, is not a simple concept – it is not a 'unitary thing' that can be 'allowed for' very easily. No clear distinction can be made between influences external to the organism and those within. Moreover, special attention must be given to influences, such as in the prenatal environment, which can mediate transgenerational effects often attributed to 'purely genetic' causes. The environment becomes increasingly complex the further one gets from the primary action of a single gene in time and space and the more genes there are involved in contributing to any aspect of phenotype.

This dynamic picture, the continual compounding of variation at all levels, is the stuff out of which individual adaptability comes. The biological machinery is able to maintain stable states (homeostasis) in the face of disturbance precisely because it is so complex and its inherent variability is in continual interaction with the environment.

Most aspects of an individual that interest us at the level of the function of that individual in society are the result of the expression and interaction during development of very large numbers of genes. Such traits are said to be polygenic. The ramifying effects of any one gene in such a group will have variable effects on the expression of all other genes in the group. For example, there may be restriction on the degrees of freedom of expression of genes that become active later in development. This applies particularly to as complex an organ as the brain which will be influenced by a large set of genes. The resulting complexity of the environment as a variable makes for developmental situations in which there is a fantastically complex matrix of interactions. The picture that Professor Rose has sketched of the extreme state dependence of the brain is a reflection of this (see chapter 7).

In this view, then, the genotype is seen as continually contributing to its own environment. This curious logical relationship, in development, of genotype and environment results in an indeterminacy wherein, to consider the extremes, variations in phenotype may correspond to the same genotype and different genotypes result in identical phenotypes. Considerations such as these have highlighted the need for a new language to deal with these epigenetic processes. For example, Waddington has put forward the notion of canalization of development. Briefly, the analogy is of a ball (the developing organism) rolling down an inclined plane, one axis of which represents time and the surface of which is composed of hills and valleys. The phenotype of the organism at any one time is defined by the position of the ball on the surface. As the ball proceeds down a valley it cannot be easily diverted from its path – this reflects the stability of many developmental processes. Moreover, when considering alternate pathways of development it can be seen that once it has entered any given valley its future path, although still not determined, has certain restrictions placed upon it. Thus the environment–genotype reticulum is considered as a complex topographic surface. Once circumstances have directed development along a certain pathway a simple modification of environment subsequently may not evoke expected pathways of development.

All this may serve to illustrate that talk of genetic versus environmental causes in relation to any aspect of phenotype represents a serious misunderstanding of the nature of development and the significance of biological diversity. This misconception possibly has its root in the carryover of the notion of cause, with its connotation of linearity and inevitability, from the physical sciences. (Although there is not scope here we should perhaps begin looking to the underlying rationale and structure of explanations in relation to problems such as this book is dealing with. This is one area of philosophical investigation which could have great practical importance.) The important point is that all aspects of an organism should be thought of as 100 per cent genetic but not 100 per cent determined. There is a subtle but important distinction which must be made between the notion of genetic origins and genetic causes. What, after all, is a 'genetic cause'? How useful is an analysis in such terms? If we are to get information relevant to the educational sphere we must avoid asking vacuous questions.

Mathematical dissection of variability

All this takes us a long way from naïve ideas about additive genetic versus environmental effects on the translation of genotype into actuality. In light of the above, what do measures of heritability tell us? Such mathematical procedures partition the observed phenotypic variability into genetic and environmental variability. A measure of variability called variance is used and the heritability is itself a ratio of genotypic and phenotypic variances. But one of the assumptions on which the mathematical method rests is that there shall be no interaction between the components of the variance. Thus the environmental variance is, by definition, due only to environmental factors that act independently of genotype. The genotype variance component is thus a 'black box' to which important environmental factors contribute in so far as they interact with genotype during development or are influenced themselves by genotype. Appreciation of this is essential in evaluating what a heritability measure means. Values derived from various studies give a heritability for IQ of 0·8 with environmental variance, as defined above, responsible for only 20 per cent of the total variance; this does not mean that environmental variables, as commonly understood, are only one-fourth as important as genetic variables. Furthermore, even if a general interaction component is allowed for in the calculations, or if the genetic variance is subdivided to estimate, for example, gene–gene interactions, such estimates turn out to be quite small and are assessed as relatively unimportant. This reflects the inadequacy of simple mathematical tools for exploration of this process of development. Given the logical relationship of genotype and environment and the indeterminacy of relations between them, one would not expect to express 'interaction' as a separate term. Can one really identify environmental factors the effects of which are independent of any previously occurring genetic event? Although algebra may of itself be faultless it can conceal assumptions which do not relate to reality. New mathematical tools are needed. Not that the analysis does not have its uses, but it manifestly cannot give very meaningful information in the study of complex developmental processes which extend over time.

Nevertheless, there are those who, claiming to be 'interactionists', continue to structure arguments about IQ differences in terms of their 'partial genetic causation'. Such platitudes should not be taken too seriously. Their importance is not to be justified by any contrast with extreme environmental positions which do not reflect modern thinking. We know that people vary genetically; our focus must be upon the *process* whereby an individual's unique genetic endowment is influenced by and contributes to his environment.

The importance of genetic research

Of late there has been much agitation for more research to look at genetic origins of individual differences in intellectual ability. What can genetics tell us?

The basic tool of animal genetics is the breeding experiment and the backcross, i.e. observing the results of mating between offspring and parents is often crucial to testing genetic hypothesis. Such approaches are obviously not possible in the human situation. Results must then be obtained by alternative approaches. One such approach is that of analysing the interitance pattern of traits amongst relatives. But while this can yield valuable information about very simple single gene traits it cannot deal unequivocally with even slightly more complex situations. Furthermore, we cannot, in a species such as man, distinguish between the ramifying effects of one single gene and the effects of other genes having a close proximity on the chromosome. So, without being able to make even this fundamental distinction, talk of searching for single genes with possible effects on specific abilities and erecting and testing 'hard' genetic hypotheses is seen to be misguided at the very least. The relevance of such conditions as phenylketonuria (PKU) to the problem before us here is questionable. It is an interesting biochemical finding, and important clinically, that this type of idiocy can be treated by dietary compensation. But this rare, obviously pathological, condition provides no model which shows how genetic data can be evaluated for complex developing characters involving very large numbers of genes. Although we can get data on the frequency of certain single genes in various populations, such data is of very limited usefulness. The frequency of any given gene in a population is not a measure of the possible functional significance of that gene in the population, because its significance has to be seen in relation to a large number of other genes and in relation to environmental circumstances. Moreover to imply that an approach similar to that adopted for PKU can, even in principle, be used to deal with the measured racial differences in IQ is misleading: firstly as to the assumption of easily getting meaningful data on arrays of genes relating to intellectual performance. But also the tacit idea of manipulation of the environment to produce the 'appropriate' phenotype, while perhaps appropriate in agriculture, represents short-sighted paternalism in the human context.

Any approach to the assessment of genetic differences in human groups must be very cautious indeed. We must face up to our ignorance. Our first priority must be an assessment of the total range of variation in the human condition. The International Biological Programme has made a start in this direction. Also special consideration must be given as to how such research bears upon the human rights and the integrity of the individual. It is in light of all this and still unresolved problems concerning methodology that we must judge the strident cries that research is being deliberately suppressed.

Attitudes to diversity

The whole question of genetic uniqueness and the assessment of the degree of variation in human populations often provokes comment about the great danger of misinterpretation and subsequent political misuse. As a consequence it is sometimes suggested that genetic variability be dismissed, *a priori*, as being of any significance in the social context. Instead, attention is drawn to conceptions of culture which preclude biological considerations. Such a view I believe is sterile, but, more importantly, gives the very dangerous impression that recognition of any genetic difference among members of the human species necessarily implies inevitable distinctions, that are judged on an axis of superiority–inferiority. To be afraid that recognition of biological differences would be used to bolster racist doctrines is to shrug off our responsibility as scientists and lend weight to the very positions we believe are biologically untenable. There is a need in our society (and in others) to hammer home again and again the notion that variability is a biological resource to be valued. This is not a prescription for social inaction nor for limp, paternalistic approaches to group variation (the difficulty in writing about individual differences is that both interpretations are made). Let us now briefly consider the biological significance of variation.

What we have said so far is that each individual of a sexually reproducing species will be genetically unique. Further, this genetic uniqueness will, in an indeterminate way, contribute to the phenotypic uniqueness of the individual. This is not to adopt an extreme empiricist point of view and say that any genotype given an appropriate development can realize any phenotype. There will be constraints; but once again they will be largely indeterminate. Each individual, in a sense, is a genetic experiment; we cannot predict what his or her potential attainments may be. The idea of a 'normal' man presents a serious semantic problem. For, unfortunately, it has connotations of a set genetic constitution against which variants are assessed as abnormal. This is the typological approach referred to in the introduction. But really what 'normal' refers to is a range of adaptation potential within the species genepool. You can see that we have moved on here to a discussion of biological phenomena at a higher level of organization, viz. that of whole populations and species genepools. How is it that whole populations change over time, and how is genetic variation involved here?

Evolution is, given certain conditions, a logical necessity and any serious objections to it as a process, on the ground that it is teleological, for example, have long since been dealt with. One of the important mechanisms by which such change occurs is natural selection, the differential contribution of individuals to subsequent generations. Phenotypic variation is a necessity for such a process and in general terms genetic variability in a population can be said to be an advantage, biologically, in that it will allow for adaptive changes to be made in the face of changes in the environment.

Various balancing mechanisms for maintaining variability in a genepool have now been theoretically and empirically demonstrated. The fact that such mechanisms have themselves evolved underscores the belief that such variation is of advantage to the population. It is in the light of such knowledge that the notion of a fixed species and of an ideal Platonic archetype against which variation is judged became completely untenable.

At the population level, the significant thing about polygenic systems is that they are tremendously important for maintaining genetic variability. Furthermore, canalization, which we saw as an important idea when considering the control of these polygenic systems during development, has consequences at this level too. If an environmental stimulus, applied at some critical period during development, results in a change in the original topography of the environment–genotype reticulum, a new set of characters will result. Genes which before did not contribute to the phenotype are now available for selection. This can result in the evolution of a new, canalized developmental scheme which may ultimately be expressed without the environmental stimulus. This phenomenon is known as genetic assimilation. The important point here is that phenotypic variability with previously hidden genetic origins can be expressed and made available for selection. The developmental epigenetic viewpoint provides important insights into how a group norm can be shifted in evolution, despite an apparent lack of potential.

As pointed out above natural selection is an important mechanism of evolution as a consequence of which different genotypes contribute differentially to successive generations. But the level at which selection works is that of the individual, of the phenotype. This simple fact has important consequences that are often overlooked. Selection does not pay attention to all the characteristics of a phenotype but only to the number of offspring that contribute to the next generation. This concept of fitness, important in evolutionary thinking, does not correspond in any simple way to the ensemble of phenotypes in a population. For many parameters will interact to contribute to biological fitness and different phenotypes, can, as it were, achieve the same score in different ways. As we saw earlier there is a similar indeterminacy in going from the phenotype through complex epigenetic processes back to the genotype. It is important to see that fitness does not relate to any gene or group of genes for a particular character. Variation from an evolutionary point of view is an essential characteristic of living evolving systems, but fitness, an evolutionary evaluation of variation, bears no direct relationship to the genes themselves. This of course overlooks pathologies, for example single gene effects of the PKU type, but these are not relevant to the case in hand. Here we are interested in the range of variations having their origins in complex polygenic systems. Our concern is with the range of existing genetic variation in the species.

Genetic alternatives maintained in a genepool are called genetic polymorphisms. They may or may not contribute to phenotypic differences which may or may not contribute to biological fitness. An obvious example is the existence, in our own species of two alternate forms, male and female. Attempts to compare them in terms of biological fitness, i.e. of the numbers of offspring they leave, is patently ridiculous. An important point emerges here; biological fitness is a technical terms and bears little relation to value systems as we know them. Patriarchy and matriarchy are values of societies not derivable from considerations of biological fitness. There has been amongst biologists and philosophers continuing debate as to whether a system of ethics can be objectively derived from evolutionary criteria. Some maintain that values can indeed be derived by, and treated with, the methods we apply to empirical knowledge, and they see the ethical systems of mankind as biological adaptations. Others say we must go beyond this to grapple with the is–ought question, to come to grips with the moral question of goodness and the problem of aesthetics. But, whichever side you come down on, social fitness of human phenotypes is not necessarily correlated with biological fitness and certainly there is no direct relation to any constellation of genes maintained in the population. Such confusion reigned supreme in Victorian times when crude notions of social Darwinism were embraced by many. The issue of race, as we know it, possibly has its roots here, with the implicit assumption that observed and inferred differences reflect some basic inferiority.

The emergence of man as a species exerting more and more control over his own environment has undoubtedly led to new evolutionary situations which we little understand. This has made for certain assertions about our evolutionary future. One such prediction has been the emergence of a stable, inevitable, genetic meritocracy. The argument goes that equality of opportunity will highlight genetic variability being expressed in relation to those characters by which society assorts individuals. This, along with assortative mating (i.e. like with like) for characters such as IQ, will result in a society that is characterized by a rigid, stable, genetically determined class system.

The first thing to note about such a gloomy prediction is that it assumes the fallacy 'equality of opportunity means environmental homogeneity'. It is my contention that we can be, must be, more imaginative than that. Secondly, it assumes that characters such as IQ are going to remain or become increasingly important to future societies. The differentiation of society will require individuals with all sorts of adaptability for dealing with situations with large elements of uncertainty; for roles will no longer always pre-exist the individuals required to fill them. IQ is not a particularly inspired measure of what our future needs may be, either in this regard or in regard to aesthetic and creative talents needed to sustain such a society. Furthermore, the meritocracy argument overlooks the existence of genetic mechanisms that can counter such stabilizing tendencies by contributing new heterogeneity to each generation and facilitating social mobility. Furthermore, that we should value each niche of society equally will become an imperative if a stable state in an increasingly complex technocratic society is to be maintained.

This brief excursion into evolutionary biology is meant to highlight, not precisely what we know about the evolution of mankind, which is very little, but to show that this perspective renders invalid simplistic assertions of genetic inferiority of various subsections. Subpopulations of the human species, races if you will, differ according to certain marker genes, but the further partitioning of genetic resources is very flexible. No subpopulation (or social class) can be said to adequately represent the whole, and every substantial part will have sufficient genetic resources to accomplish new phenotypic norms under the appropriate conditions of selection. To reiterate: genetic variability is a biological resource to be valued not judged according to ill-conceived notions of superiority–inferiority.

Genetic diversity – then what?

Faced with genetic uniqueness, usually two approaches are advocated. Either we should attempt to equalize phenotypes as much as possible over a wide range of environmental conditions or we should, by equalizing opportunity, provide a uniform environment. Here is the dilemma of the deeply rooted egalitarian values of liberal societies. Which way do we go? But these, biologically, are false alternatives.

The first option is practically unrealizable in large measure and harks back to the typological modes of thinking, that, I hope, we have shown are inappropriate. Furthermore, it is biologically and socially undesirable. The arrogance implicit in assuming we know what characteristics will be of value in the future makes for unfortunate paternalistic and weak liberal approaches to human variability.

The second possibility is more interesting. A corollory of optimizing each individual's development is that, in any one set of environmental circumstances, individual genetic differences will tend to be exaggerated. Now obviously, this is a caricature of what happens in say, an educational situation, for how uniform is the experience, given differences in family background? But in general terms it can be said to be true. The curious thing is that in some quarters arguments sustained in this way have been warmly embraced as in some way bolstering up ideas of inevitable, refractory, genetic causes of observed IQ differences. This is a *non sequitur* as by now the reader will appreciate. Besides the connotations of immutable superior–inferior distinctions there is another fallacy. Do efforts to promote social progress necessarily imply homogeneity of environment? Any arguments based upon improved nutritional status and medical care (these must be improved in large sectors of western society and the Third World) which remove crudely imposed environmental deficits do not provide a case for environmental homogeneity. Predictions based upon such crude notions of equalizing environment must be dismissed as unduly pessimistic and serve only to deter us from adopting a social and political stance for bold and innovative approaches to the problem.

The difficulty here is that deeply rooted ideas about equality of opportunity refer, by implication, to the present educational milieu which is still very hierarchical, authoritarian and mirrors the value judgements made on various occupational niches by present society. This is hardly likely to be the best means for investing our potential in the future.

The poverty implicit in an approach wherein a measurement of IQ determines the subsequent parcelling out of the appropriate educational experience drawn from a few alternatives does not do us proud. We surely must extend any ideas about individualized education beyond that if we are to capitalize on the wealth of genetic variability we have. One approach might be to look at the possibility of constructing very diffuse matrices of educational experiences in which individuals assort and move at their own speed. Many possibilities are before us, but we need to plan for a future wherein society will be throwing up many more niches which will need to be occupied by people assured of their essential integrity and humanity. We must go beyond essentially static, hierarchically oriented social situations to one wherein the richness of individual differences is seen as a treasured resource and no one group is assumed to have a monopoly on the genetic variation essential for future evolution.

I feel sure that the psychometric approach is neither constructive nor imaginative and will in years to come be seen as foolish, inappropriate and without real biological foundation – a misleading perspective on the future of mankind, which is, after all, an important aspect of education.

Further reading

Several works on the present state of biological knowledge are included which, although not directly relevant to the present issue, may provide the interested reader with a more appropriate perspective.

Bernard Campbell, *Human Evolution*, Heinemann Educational, 1963.

T. Dobzhansky, *Mankind Evolving*, Yale University Press, 1962.

H. J. Eysenck, *Race, Intelligence and Education*, Temple Smith, 1971.

Jerry Hirsch (ed.), *Behavior: Genetics Analysis*, McGraw-Hill, 1967.

L. Hogben, *Nature and Nurture*, rev. edn., Allen & Unwin, 1939.

L. Hogben, *Statistical Theory*, Norton, 1968.

Journal of the History of Biology, vol. 2, no. 1, 1969. Entire issue is devoted to explanation in biology.

James C. King, *The Biology of Race*, Harcourt, Brace & World, 1971.

A. Koestler and J. Smythies, *Beyond Reductionism*, Hutchinson, 1969. Note particularly the essay by C. H. Waddington.

Watson Laetsch (ed.), *The Biological Perspective: Introductory Readings*, Little, Brown, 1969.

I. M. Lerner, *Heredity, Evolution and Society*, Freeman, 1968.

J. A. Moore (ed.), *Ideas in Modern Biology*, 16th International Congress in Zoology, 1963.

M. Ashley-Montagu, *Culture, Man's Adaptive Dimension*, Oxford University Press, 1962.

D. N. Robinson, *Heredity and Achievement*, Oxford University Press, 1970.

C. H. Waddington, *New Patterns in Genetics and Development*, Columbia University Press, 1967.

A. Roe and G. G. Simpson (eds.), *Behavior and Evolution*, Yale University Press, 1958.

7 Environmental Effects on Brain and Behaviour
Steven Rose

Steven Rose is Professor of Biology at The Open University. He obtained a double first in biochemistry at Cambridge and a Ph.D at the Maudsley Institute of Psychiatry, London. He held both the Beit Memorial and Guinness Fellowships at Oxford and worked at Imperial College, London, for five years. He researches into the biochemical bases of brain function.

Environmental effects on brain and behaviour

So far this book has considered various aspects of the performance of individuals and groups, as measured by IQ tests and other ways, and also what can reasonably be concluded concerning the genetics of such performance. In all these accounts, there has been remarkably little said about the organ of the body which is above all responsible for behavioural performance and that complex set of attributes defined as intelligence – the brain. Rather, the individual and his brain have been treated somewhat as a black box, with particular sorts of output. It is the purpose of this chapter to examine the structural, physiological and biochemical features of the brain which are relevant to an understanding of those aspects of its performance subsumed under the general title of intelligence, and to consider the evidence relating to factors which can influence this performance. Much of this evidence comes, inevitably, from animal studies, and it must always be borne in mind when evaluating it that extrapolation upwards to men is beset with difficulties. It is no more desirable socially, nor sound scientifically, to be rattomorphic or chimpomorphic about humans, than it is to be anthropomorphic about animal behaviour.

Brain structure and performance

A nervous system may be defined as an organized constellation of cells (*neurons*), specialized for the repeated conduction of an excited state from receptor cells or from other neurons to effectors or to other neurons. It is essentially a device, first, for the reception of sensory information, second, for its processing, storage and comparison with past information, and third, for making decisions about how to act, on the basis of all this data, and instructing the effector organs of the body accordingly. The greater the storage and processing capacity of the system, the more effective it will be at these tasks.

Effectively to perform them, the nervous system needs to show two complementary features. If it is to be able to function at all, it needs to have a set of built-in, programmed responses. Particular patterns of sensory input must result in certain predictable outputs. A tap on the knee must result in a knee jerk, pain in a limb to its removal from the stimulus. These are examples of the *specificity* of the nervous system. In addition, however, it must have the capacity to respond to new information in a novel way. If a particular type of food tastes bad, it should not be eaten again. If a given type of activity produces reward, it should be repeated. This represents the *plasticity* of the system. Specificity is the genetically programmed, invariant response of the system, whilst plasticity is the learned result of experience. For any species, behaviour is a result of the sum of – often the tension between – the plasticity and specificity of its nervous system.

In all vertebrates, and particularly in mammals, the largest group of specialized cells performing the functions of the nervous system is in the brain, and an examination of human performance must, therefore, begin with examination of the human brain. The adult human brain weighs 1300 to 1500 grammes – heavier than most organs in the body. In appearance it is dominated by the large, convoluted, walnut-like masses of the cerebral hemispheres, which fold over and bury beneath them practically all other structures. The hemispheres themselves consists of a skin, three to four millimetres thick, of 'grey matter' above an internal core of 'white matter'. This thin skin is the cerebral cortex, and is densely packed with nerve cells. The white matter beneath gains its characteristic colour from the high concentrations of a fatty substance, myelin, which forms the insulating sheath round the many nerves which run to and from the neurons of the cerebral cortex. Of the many structures of the brain, it is the cerebral cortex which is most concerned with conscious behaviour, the processing and analysis of incoming sensory information, and decisions as to appropriate motor responses. The cortex is the most plastic part of the brain, concerned with learning, memory and the coding of the experience of the individual.

The evolution of intelligence and consciousness

Thus, in an examination of factors which affect intelligence and performance, we may guess that the area most relevant to our study will be the cerebral cortex, although clearly many other regions of brain and body will have a part to play. Deficiencies in sensory input or motor output will affect performance. So will factors relating to attention and alertness, controlled by regions lower in the brain. One way of approaching the question of human intelligence may then be to ask how the human brain resembles, or differs from, the brains of other animals. The human brain is not, by any means, the largest of all. Elephants, dolphins and whales for example, all have heavier brains, but these animals also all weigh a great deal more than humans. It is a reasonable postulate that the more body cells there are, the more brain cells will be needed to control them. So a fairer estimate of brain weights is to relate them directly to body weights. When this is done the human brain ranks amongst the highest. But the differences are not dramatic. Clever as dolphins are humans are a great deal cleverer, and this is not reflected in any massive difference in the ratios. More striking perhaps, is the difference in the size of the cerebral cortex of humans compared to other species. In the evolutionary path to man, the development of the brain is characterized by a progressively greater dominance of the cortex.

The increase in cortex size, however, is not so striking as the increase in learning capacity and plasticity of the brain, which is the unique feature of humans. Fishes can learn simple avoidance and discrimination responses, pigeons to count, rats to run mazes or press levers for rewards, dogs to round up sheep and perform other complex tricks, dolphins to communicate with humans, chimpanzees to use simple tools, generalize and even, in some recent experiments, to construct sentences and use language symbols. All these are examples of brain plasticity: they are not inherited, but learned skills. In general it may be argued that, in the evolutionary path to man, there is an increase in plasticity, and a relative diminution in specificity. The specific 'innate' responses of organisms with smaller brains, from the fluttering of a moth towards a flame to the complex community relationships of ants and bees, are all based on specificity rather than plasticity. Their responses, at their most complex, as in the bees capacity to make maps of the external world and to communicate them to the fellow members of its hive, are striking. The learning capacity of the bee on the other hand, is almost negligible. But with the human there is a quantum jump in performance, intelligence and consciousness.

What structural feature does this performance depend upon? In part, clearly on non-brain features. The human hand is capable of more complex manipulations than that of the ape. The structure of the human vocal and auditory systems is better adapted to making and interpreting complex sounds. But there must also be a combination of brain features which enable these advantages to be exploited. These features are clearly not just brain size, or cell number, or even cortex size. It seems probable that there is another crucial aspect of the organization of the brain which is relevant to performance. This is its connectivity.

Within the nervous system, neurons are connected to one another in such a way that signals arriving, say, from sense receptors, can be transmitted down other nerves to the effector organs, like muscles. The complexity of the instructions that can ultimately be transmitted depends on how much information, about present and past events, can be collated and computed before the final message is transmitted. Without going into the microscopic structure of the brain system, it is fairly obvious that the more neurons that are involved in the making of any decision, the more complex are their interactions and the more information the final message will be based upon.

The point at which one neuron makes contact with another,

across which information can pass between them is called the *synapse*, and each neuron makes many synapses. The synapses are thus the main information processing devices of the nervous system, for they enable a whole set of incoming data to be compared, collated and either acted upon or not. The synapse is the decision point of the brain; it has been likened to the yes/no gate of a computer. The capacity of the brain to store information depends not only on the number of its cells, but the number of synapses between them, which determines the number of possible interactions. And there is some evidence that this number is larger in men and monkeys than other animals. The neurons of the rat cerebral cortex may each make only some 10^2–10^4 synapses; in monkeys and some regions of the human cerebral cortex the figure may be an order of magnitude higher, from 10^4–10^5. There are estimated to be 10^{10} neurons in the human cerebral cortex, perhaps 10^{14} synapses. This is 30,000 times as many synapses as there are humans on earth, and we may postulate that the specifically human aspects of intelligence and performance are indeed some function of the number of neurons and their connectivity. To ask the question 'what determines intelligence?' can then be rephrased in neurobiological terms as 'what factors decide or influence neuronal cell number and connectivity in the brain?'

But to ask the question only in these terms misses one highly significant point. In listing the features which distinguish man from the apes, we did not include one of the crucial ones; his capacity to live and communicate in social groups. Because of this capacity to communicate with his fellows, first in words, later in the more permanent form of writing, the information available to the human, even from early days (30,000–100,000 years ago) surpassed by orders of magnitude that available to his evolutionary neighbours. Storage and transfer of information, outside the brain, became possible, and the experiences of one individual could be transferred to another, even across generations. Hence the social evolution of man, which is so important in understanding the present situation, could begin. The difference in performance, and doubtless intelligence, of the men of today from the earliest of *Homo sapiens* is enormous, precisely because of this factor. The cranial capacity of the early humans does not suggest they had a greatly different brain size from ours, yet the range of their activities was much more limited. Yet the 30,000–100,000 years that have elapsed since the early days of man is far too short a time for major genetically derived evolutionary changes to have occurred. To ignore the social environment in which man operates, to ignore the fact of man becoming, and to postulate any model of brain function as an absolute outside this social environment, is to be guilty of an error of cardinal significance.

The development of the brain

So far we have been concerned with the performance of the human brain compared with that of the animal. But what distinguishes the performance of one human brain from that of another? This is clearly the key question with which this book is concerned. One obvious distinction might appear to be brain size. Some people have heavier brains than others. But when expressed in terms of the brain weight to body weight ratio, there is surprisingly little difference between sexes, races or individuals. Post-mortem examination of Einstein's and Lenin's brains revealed no significant differences in terms of weight ratio or obvious structures from that of the average human. The performance differences must depend on micro-structural differences at the level of synaptic interactions in the living brain which no post-mortem study of pickled sections can reveal. Whilst one thus cannot detect in individual adults the cause of differences in performance, an examination of the pathways of development of the brain during infancy may help to do so.

Amongst all the organs of the body, the brain is unique in that the neurons are a non-dividing cell population. Each of us is born with very nearly his full complement of neurons, a very few more being formed in the first months after birth, and the brain at birth is closer to its adult size than any other organ of the body. It comprises 10 per cent of body weight compared with 2 per cent of the adult. By six months old the brain is 50 per cent of its adult weight, at a year 60 per cent, at five years 90 per cent, at ten years 95 per cent. The nine pre-natal months are thus a period of enormous brain cell growth, and most of the key brain structures are established at birth. The developments after birth are of three kinds. First, the proliferation in the brain of a type of cell other than the neurons (glial cells) which have a predominantly supportive role in brain function. Second, the development of the synaptic connections of the neurons. Third, the laying down of the myelin sheaths of the nerves, which enhances their function, and hence the development of the characteristic brain 'white matter'. There is a complex but ordered pattern to this development which is the expression of the genetic programme of the organism in interaction with the environment.

It is during this pre- and post-natal development that the interplay of specificity and plasticity becomes of such major significance for the organism, and it may be as well to clear up here some of the misconceptions that abound in this area. The genetic programme of the individual is an expression of the DNA content (the genes) of the egg and sperm from which he develops. But this genetic programme can never be expressed without an environment in which the expression is to occur. If the environment is inadequate, the individual simply dies. From the first moment of cell fusion, there is an interplay between the genes and the environment of the most complex and interactive kind. During development, the pattern expressed on any gene becomes part of the environment of all the other genes. Even marginal differences in the external environment may induce a variety of changes in the nature and quantity of the protein being expressed upon the genes. It has been observed in a certain mosquito, for example, that a phenotypic female can be produced from a genotypic male simply by exposing the organism during development to an elevated temperature (29°C). In female rats, implantation of testosterone, a male sex hormone, into the brain of the young animal can produce male sexual behaviour in the adult. To attempt to parcel out hereditary and environmental influences during such developmental sequences is meaningless.

The interaction of genetic programme and environment is not really difficult to understand, yet it has been made the subject of vast oversimplifications. Thus in a recent book one psychologist referred, in apparent good faith, to the existence of 'high

IQ genes'. Brain performance depends on structural interactions, which in turn depend on changes in a large number of interacting biochemical systems. A recent analysis shows a considerably higher percentage of the brain's genome – its DNA complement – is switched on to the manufacture of protein than in any other organ. The genes of the brain are producing more different types of protein – perhaps 30,000 different ones in all – than any other part of the body. A large number of these proteins presumably play a role in specifying brain structures and hence performance. Only in very rare cases – the one or two genetic diseases in which one of the genes is absent, as in phenylketonuria, for example – is it possible to specify the effect of the absence of a particular gene in relationship to function. As Bodmer points out, phenylketonuric children, if untreated, suffer considerable brain and performance damage. The gene which prevents phenylketonuria, because it produces a key enzyme which helps metabolize the substance phenylalanine, is present in normal children, and if they are compared with the phenylketonuric children, it is obviously by this definition a 'high IQ gene'.

But even here the child can be spared brain damage if its environment is modified at birth. If for example it is fed a diet containing no phenylalanine, it will develop practically normally. Hence the environment has 'triumphed' over the genetic deficiency of the individual. The fact is that, unlike a simple trait like eye or hair colour, brain performance is profoundly complex. Not only does it depend on a very large number of genes, but it displays a plasticity such that if one or several of these genes is modified the performance of the others is affected so as to tend to compensate. To talk of 'high IQ genes', or to try to disentangle the genetic programme from the environment in which it is expressed is both disingenuous and misleading.

What can be studied are the differences between the developmental pathway which is taken in a 'normal' environment – that is one varying over rather small limits – and one in which quite substantial changes outside these limits have occurred. The problem of interpretation, however, remains; for plasticity, that is the capacity for continued interaction with the environment, is itself programmed into the genetic specification of the human.

If the brain were rigidly and deterministically specified it would be useless as a brain.

Certain parts of the system are, however, laid down fairly rigidly. A supreme example is the pattern of pathways and interconnections of the visual system, from the retina of the eye via various intermediate neurons to the visual region of the cerebral cortex. If these pathways were not specifically programmed the organism would not be able to receive and analyse visual input and would be disadvantaged. What are not specified, however, it would seem, are the connections of the visual analyser cells of the cortex themselves. Much of this region of the brain is relatively undeveloped at birth; electrical activity develops in it only slowly and myelination is also retarded compared with other brain regions. The newborn child may be able to see, but it is doubtful if it can analyse or understand what it sees. The development of this capacity to analyse depends on subsequent interaction of the environment with the brain and can indeed be modified, as will become apparent subsequently.

Environmental effects upon brain structure and performance

I now turn to an examination of the way in which such modifications of the environment can and do affect both brain structure and brain performance. One of the severest environmental alterations which can be provided, short of actually physically damaging the brain, is to deprive the organism of food. When this is done in the adult the body weight declines sharply, but the brain weight is relatively unimpaired. The biochemical defence mechanisms of the body protect the brain against being used as a food reserve practically until death from starvation. The body sacrifices almost every other organ in preference to the brain.

But in infancy and during brain development the situation is different. Work on experimental animals, particularly that of Dobbing, has shown that if they are malnourished or undernourished for periods during weaning, or in the period just following, which are the times of rapid brain growth, then, not only will body weight be dramatically affected, but brain growth itself will also be retarded. Even if the animal is subsequently fed freely, with as much food as it can eat, the brain growth may never catch up. During infancy in the rat or the pig for example, there are thus certain sensitive periods during which, if brain growth is impaired by malnourishment, the effect will last for the lifetime of the individual. In the case of the rat, this malnourishment can be achieved by the relatively simple method of taking two litters of pups born the same day, mixing them and returning a few – three say – to one mother and the rest – which may be up to fifteen or twenty – to the other. The 'large family' pups will have permanent deficits in brain weight which cannot be remedied after weaning, however good the diet they are on. Under these circumstances, the brain weight/body weight ratio stays permanently outside the normal limits.

In addition, there is a deficiency in the level of brain DNA (which is presumably a measure of brain cell number) which is never retrieved in these undernourished animals. Such deprivations will also result in performance deficits. Baird and her colleagues have shown that rats malnourished from conception, birth, or weaning, made more errors when tested on a Hebb-Williams maze than controls, either during their period of malnourishment or after a period of five weeks of rehabilitation.

How far are these observations relevant to the human situation? Rats are different from humans, not only in the obvious ways, but also in terms of the state of the development of their brain at birth. Compared with humans, rats are born relatively retarded. A baby rat is blind and naked. Biochemical studies show that a major spurt in its brain development occurs in the first two post-natal weeks, when a large variety of structural and biochemical systems rapidly mature, including the glial cells, synapses, and myelin formation. This then forms a critical developmental period for the animal. The human situation is different. Relatively more of the human infant's brain development has occurred prior to birth, and Dobbing has suggested there are two critical periods in human brain development, one pre-natally, the other the period of glial proliferation and myelination over the first eighteen months to two years post-natally.

To extrapolate from rats to humans is neither easy nor

sensible. But in many parts of the world, child malnourishment occurs, and its scale is terrifying. The key problem is protein deficiency, and it has been estimated that more than half of all the children in the world are 'at risk' from effects of protein deficiency, manifested in the most serious cases as the disease Kwashiorkor – but even in less serious cases the deficiency is on a scale likely to cause irrevocable impairment of brain development. This is the most serious nutritional disease in the world, affecting most strongly the children in the lower socio-economic groups. Three hundred and fifty million children under the age of six are probably thus affected.

The relationship of brain-weight to nutritional status is revealed clearly by studies in Latin America, which show that severely undernourished children have a reduced head circumference compared either to high socio-economic groups within their own country, or to American or western European children. Coupled with the small head size goes small body stature, increased age of puberty, and a variety of other features as well, not directly relating to brain performance. Nonetheless, even when the calculations are made on an estimated brain-weight to body-weight ratio, such undernourished children, or adults who have been undernourished in childhood, show deficits. We may expect to find that the critical periods for malnourishment in humans fall both pre-natally and in the first post-natal years.

Similarly, performance deficits occur in undernourished children. Cravioto and his group have studied such children in Mexico, and shown that compared with the children of rich parents in the same country, they have reduced sensori-motor skills and capacity to associate across modalities – pretty basic performance deficits which clearly will manifest themselves in such measures as the IQ tests.

Children in America, Britain or western Europe are unlikely to be as undernourished at those in Latin America or India. Nonetheless, longitudinal studies on British children have shown that small body size tends to be correlated with low socio-economic status and large family size. That is a polite way of saying that they come from a family with not much money to spend on food. Once again the critical question is how far actual performance is affected by such poverty and malnutrition. Whilst it becomes extremely difficult to sort out many of the other related variables in such studies, the longitudinal studies suggest that low IQ scores are also associated with low socio-economic background of the parents, large family size and poor maternal state of health during pregnancy. Pasamanick and his associates have found that deficiencies of maternal diet associated with low income can produce complications of pregnancy and parturition followed by intellectual retardation and behavioural disorders in the children. Harrell and his co-workers added nutritional supplements to the otherwise deficient diets of pregnant women, and found significantly raised IQs in their children at ages two, three and four compared with the children of mothers without the supplements. Whilst gross undernourishment is infrequent in Britain, malnourishment and the diseases of deficiency are disturbingly frequent. It would seem probable, therefore, that a considerable number of performance deficits in children from deprived backgrounds are associated with long term impoverishment of this sort.

In terms of human welfare and politics such effects are surely of great significance, because the obvious way of remedying them lies to hand. Elaborate arguments as to genetic effects on intelligence, which present a prescription for social and political inaction because they are apparently at best irremediable, can only hinder the eradication of this monstrous situation.

However, it may be argued that the discussion does not come to grips with the issues of whether less extreme environmental situations can produce effects on brain performance. In this context, the work of groups studying the effect of changed environments on brain chemistry and performance is of relevance. Some of the most detailed of these have been made by the group of Bennett, Krech and Rosenzweig, who rear littermate rats from birth in one of two types of condition. One type 'environmentally impoverished', where, although fed enough, the animals are caged individually in conditions of low sensory stimulation of light or sound, out of sight of their fellows. Their handling is reduced to a minimum, or avoided altogether. The comparison group is reared in an 'enriched environment' living in a communal cage, and handled often, with plenty of 'toys' to play with and objects to explore. At the end of some weeks in either of these two conditions, the animals are killed and certain characteristics of their brain chemistry examined. Environmentally enriched animals are found to have a thicker cerebral cortex than the impoverished ones, and certain brain enzymes, notably those associated with synaptic transmission, such as acetylcholinesterase, alter in concentration. As little as one hour a day of the enriched experience is enough to cause measurable differences in these parameters. Very recently, this group also claimed that there are significant differences in the size of the synapses in the cerebral cortex of these animals when viewed electron-microscopically.

These changes relate to brain structure and chemistry. The question of performance changes remains to be described. These have been studied to some extent by the Bennett, Krech and Rosenzweig group, who do find performance differences between their two groups of animals.

The recent experiments of Blakemore and Cooper, and of Hirsch and Spinnelli, show that quite basic brain mechanisms are environmentally modifiable. In the Hirsch and Spinelli experiments, kittens were reared under conditions of controlled visual experience, so that one eye was exposed only to vertical lines and the other simultaneously to horizontal lines. At the conclusion of the training period, it was found that the cells of the two halves of the brain responded differently to visual inputs, each side responding best to the type of pattern that was familiar to it through the training. Early experience in these animals has directed subsequent physiological performance, even in as apparently genetically 'programmed' a response as the development of the analytical capacity of the visual system. Many other 'innate' and 'specific' behavioural responses are now known to be a good deal more plastic and modifiable, and less specific, than was once believed.

But the classic, and perhaps most relevant examples of the effects of early environmental experience on behavioural functions remain the well-known experiments of the Harlows, who showed that rearing monkeys in isolation resulted in an apparently permanent inability to form normal social, sexual or parental relationships in later life.

How far this type of experiment is related to the human experience is of course subject to all the cautions about extrapolation that have already been made. Environmental impoverishment to the extent involved in these studies, such as rearing in the dark, is obviously rare. The contrast between the impoverished and enriched environment of humans, even at its most extreme, cannot be so dramatic as this. Nonetheless, the differences are there, and all the evidence from the animal studies must lead us to suspect differences in humans in response to environmental changes as well. What can be less easily separated in the human situation are the nutritional and environmental effects, as children from impoverished environments are more likely *a priori* also to suffer nutritional disadvantages. The most relevant studies in this regard are the longitudinal observations made in Aberdeen and the experiment of Skodak and Skeels, referred to by Professor Bodmer (see chapter 5).

Transgenerational effects

The thrust of the argument so far advanced has been that environmental effects on brain performance and brain structure are profound and must always be so confounded with the genetic ones that parcelling out is meaningless. But we come now to the most important class of effects of all from the point of view of this argument, effects which, in the human situation are formally totally indistinguishable from the genetic ones, and yet are consistently ignored by those who would claim to have characterized an isolable genetic component in brain performance These effects are referred to as transgenerational.

We can discuss them first in terms of the animal experiments. It has already been pointed out that nutritional deficiencies in childhood result in permanent changes in brain chemistry for which a subsequent adequate diet does not compensate. Recent experiments have also examined the effect of these deficiencies in the second generation. If malnourished female rats are then allowed to produce offspring, what will be the effect of the maternal infantile malnourishment on the brains of the offspring? Several intriguing experiments have shown that the undernutrition of the mother is reflected in low brain weight in the offspring. The most recent results are those of

Zamenhof and his group, who earlier had shown that protein deficiency in infancy in the rat results in lowered DNA (lower cell numbers) in the adult. They then mated the females amongst these infantile undernourished animals with normal males. Amongst the offspring, whether weaned by their own mothers or by normal foster mothers, there was also a highly significant reduction in DNA content, and hence probably in cell number, in the brain.

Such effects are transmitted between one generation and the next. Yet they are not genetic but environmental. Presumably after an adequate number of generations have been adequately nourished, they are reversible. Such effects are also reflected in performance. Again the Harlows' work may be cited. Not only is social behaviour in monkeys reared under deprived conditions itself abnormal, but this abnormality reflects itself in the rearing of the animals' own infants. Thus the rearing of the second generation is affected by the childhood experiences of the previous one.

In other words, the Harlows' monkeys are demonstrating the truth of the well known observation that the sins of the parents are visited upon the children. Except that in this case – and maybe in most other cases as well – the parents themselves are more sinned against than sinning. Children reared in deprived conditions are 'at risk' in two ways. Not only are they more likely to suffer from diet-linked complications of pregnancy, but the consequences of this are more likely to ensure that they grow up and live in the same adverse conditions as their mother. Such handicaps combine to sustain the very conditions from which they are derived, so the vicious circle tends to be repeated for generation after generation. Clearly the correlations between IQs of parents and children are bound to be high, and give a superficial impression of genetic determination. To fail to appreciate this is to ensure that bad science is done and bad social policy advocated.

Conclusions

Of course, this all ought to be self-evident. Everyone knows, from their own experience, the extent to which their adult behaviour reflects their childhood experiences, and indeed, anyone with any degree of self-consciousness at all will be aware how much his own childhood experiences affect the rearing of his own children. And if we did not know it from our own experience – that non-scientific guide – we would know it from the work of the psychoanalysts. It is easier to understand and accept these effects in humans than it is in animals. And indeed, it is obvious from our experience, and should not, perhaps, need to be said, that 'enriched experience' in childhood modifies subsequent performance. It seems almost bizarre that we should have been put in a position of needing to reaffirm some of these statements.

Brain structure and chemistry determine performance, and brain structure and chemistry are themselves subtly but profoundly affected both by immediate environmental influences and by those stretching back beyond our own generation and into an indeterminate distance into the past. Brain structure and performance are affected most seriously by the world tragedy of protein starvation in the developing countries, which puts 350 million children at risk, and of malnutrition and appalling social environment in the industrial societies of Europe and the United States. In situations in which parents are expected to bring their children up in the slum conditions of the Gorbals or Notting Hill or Harlem, fed badly and overcrowded at home, deplorably educated in overcrowded schools, condemned to lives in which unemployment may alternate with the alienated labour of an oppressive social order, the miracle of it all is the human capacity to triumph in spite of adversity, to succeed despite the odds. This is the key to human social evolution, to man becoming; his capacity to transform his environment positively.

What is quite intolerable is that, rather than throwing their weight behind the obvious measures which will immediately improve the environment and hence the performance of such a large proportion of humanity, certain self-styled advocates of 'pure science' – whatever that might mean – should continue to attempt to cloud the situation with a set of spurious arguments which have the effect – whatever the intention of their advocates – of justifying the present situation.

Further reading

N. Chalmers, R. Crawley and S. P. R. Rose (eds.), *The Biological Basis of Behaviour*, Harper & Row, 1971. Selected readings in neurobiology.

J. Dobbing, *Growth and Development of the Human Brain*, Saunders, 1972. A comprehensive review.

K. Oatley, *Brains and Minds*, Thames & Hudson, 1971.

S. P. R. Rose, 'Neurochemical correlates of learning and environmental change', in G. Home and R. Hinde (eds.), *Short-Term Changes in Neural Activity and Behaviour*, Cambridge University Press, 1970.

S. P. R. Rose, *The Conscious Brain*, Weidenfeld & Nicolson, 1972. An overview of neurobiology, its philosophy and social implications.

N. S. Scrumshaw and G. E. Gordon (eds.), *Malnutrition, Learning and Behaviour*, MIT Press, 1968.

From Sociology

The environment, the last of our list of central notions, is the theme here. Donald Swift argues that in just the same way as intelligence has come to be regarded as a 'thing', the environment is often taken to be a simple physical factor which can be assigned a number and added to the genes. Notions of development discussed in the last section have already warned of the misleading simplicity of this view, here Donald Swift goes a step forward and shows us that our environment is also in our heads. The environment which influences us is partly our view of the world and the same thing may look very different to different people. He goes on to outline the kind of theory of the environment we require to begin to understand development in realistic terms – a theory he finds lacking in the work of psychometricians. This theory is continuous with the biological view of development described earlier by John Hambley. He also strives to rid us of some of our ethnocentrism, pointing to a sociological view which regards varying social classes and racial groups as different but not inferior. Some of the implications of these views for compensating education programmes designed for the disadvantaged are discussed, and the 'deficit' view of such groups is argued to be an important factor in the apparent failure of such programmes.

From other directions, John Rex criticizes the 'crude empiricism' of some psychological and sociological attempts to quantify social factors, and he argues that the cult of numbers produces 'facts' without understanding. In the second part of his chapter he turns his attention to the social context of the race – IQ debate itself. He explains how in the post-war years, after the excesses of Nazi race theory, the whole question became a taboo subject in academic debate.

This was a period of concensus views which stressed similarities within societies. More recently the concensus has broken and racial and social divisions have come to the fore and are reflected in academic posturing about the origins of the differences. He sees the role of much of this academic writing as providing rationalizations for views already latent in society. These rationalizations are not peripheral, however because they provide an incentive for new levels of discrimination and division.

In the conclusion we take up some major points which have recurred throughout the book, especially those concerning the limitations of the psychometric approach to human behaviour. From this point we go on to examine educational innovation programmes like Headstart in the USA. Have they failed? The discussion of this question leads us to take a more searching look at what they were intended to do – a subject that appears not to have been considered adequately when the programmes were started. Is raising IQ a sensible social goal? Enough has been said about IQ to make it a very doubtful criterion. Beyond this there are further questions, is educational change an effective way of modifying social structure? That is what is involved if you give the poor the means of 'taking their place in the mainstream of society'. Perhaps the relationship of social structure and education operates the other way round and educational reform can only be carried out within narrow limits determined by current social organization.

8 What is the Environment?
Donald Swift

Donald Swift is Professor of Education at The Open University. Prior to this he taught in the Department of Educational Studies at Oxford University. He taught in Liverpool slum schools before becoming a sociologist. He has published in the area of social factors affecting the development of ability, and is an editor of *Sociology of Education Abstracts*.

Inadequate notions in the study of human behaviour

To the sociologist the old nature–nurture controversy is at best an interesting platform for methodological discussion – for an argument on the status of our knowledge at any point in time – but nothing very serious. It only becomes serious when this incredibly complex area of insubstantial knowledge is used as a source of 'expert knowledge' which is 'true' in a way which ignores the extreme limitations on the reliability and validity of our data. That is, when we become too presumptuous and too arrogant for genuine scientists.

Ever since I taught 'backward' children who were clearly of high intelligence I have found it difficult to understand the conventional wisdom of educational psychology about the sources of exhibited ability in school. The assumptions that were made about life in 'the humbler classes' were so unreal, not to say patronizing, as to warn one that people who know so little about the lives of their subjects, had to be taken with a large grain of salt. But that was only the natural scepticism of the practitioner towards the head-shrinkers whom we liked to see as odd, though clever for getting more money than we did for less work. I lost that approach in my first postgraduate contact with teachers in a research project. During a seminar given by such a 'head-shrinker', I was taken on one side by one of the participating teachers who explained to me that I could ignore whatever the head-shrinker said as he had been a 'conshie' in the war. Arguments like that are not going to get us very far in understanding the consequences of our attempt to manipulate the experiences of children, but they do raise an interesting point. What do we mean by scientific evaluation of evidence? Clearly, the question of whether genetic or experiential factors contribute most to exhibited levels of ability gives plenty of room for variation amongst scientists. We should resist the easy explanation that people who differ from us do so because of their own incapacities, and covert motives.

Let me concentrate, instead, on the assumption that they differ in the meanings they are prepared to attach to the numbers that come out at the end of research – for numbers have no meaning themselves – *we* have to attach meanings to them. We can start by insisting that we are doing science – with the hope that better knowledge will make us better able to improve our lot. Science is dependent upon the classical experiment as its means for verifying information. That is, the results of any piece of work are dependent upon the extent to which the researcher has obeyed the requirements of this perfect means for isolating relationships which he may reasonably, though tentatively, consider to be causal.

The problem with the classical experiment is that is can never be achieved. In research upon non-human material we can often get close enough to satisfy everybody (at least at that point in time – ten years later is often a very different situation). But our research upon humans, upon sentient matter acting with intentions, carries ethical and scientific problems which do not necessarily put us in a qualitatively different position from the researcher on non-human matter but which certainly complicates things greatly. Obviously, we could answer many of our questions better if we could copy the Egyptian King, who, wishing to find out the natural language of the human, incarcerated a hundred babies away from all experience of human communication. But, on ethical grounds we cannot interfere with the social process except in trying to do 'good' things to children; we can only experiment in one direction on the social side of the nature–nurture interaction. On the nature side we cannot experiment at all.

This leaves us with two alternatives. We can make do, either with 'after-the-event analyses' of situations we find in society (like the study of some identical twins whom we have found to have developed in separate cultures) or we try some 'beneficent' experiments (like giving special treatment to one child from a particular culture and ignoring a similar one to see, if by comparison, our interference has had a 'good' effect). Neither alternative provides an adequate vehicle for highly reliable knowledge in this area. In both these cases the level of our satisfaction with the data we arrive at will be a function of our evaluation of the *quality* of the research. This is always so, of course, but more crucially important in the non-experiment because we have no overall benchmark to determine quality. In the case of the experiment we can check whether any of its requirements are missing – if so, it is inadequate and we are left to salvage what we can. In the non-experiment we *always* have to make up our minds about the comparative weight we will attach to specific weaknesses; we are always salvaging. Most importantly, we have to balance the quality of the ideas that have been used in producing the data with the quality of data-gathering and the quality of the chosen methods for data analysis.

Argument about the quality of ideas is at the root of the debate on the nature–nurture problem. It is a problem of different research traditions and conceptual levels. The model of research in psychology is the classical experiment. In a laboratory it is often possible to get away with a piece of work which ignores the theory that has been used to suggest which bits of the social world we choose to 'measure'. Outside the laboratory it becomes impossible. In the experiment we must assume that we have controlled or randomized all possible contaminating or intervening variables, and if we have a very tightly circumscribed problem – perhaps a subject's response to crude bits of mechanical communication like flashing lights or photographs – it may not be all that difficult. It may also produce such trivial information that lack of theory does not matter. Outside the laboratory, on the other hand, where the number of possibly important variables is limitless and mostly uncontrolled, the quality of our results is strictly a function of the quality of our conceptualization (ignoring technical ineptitude). That is, only good theory makes clear why we have chosen some variables and ignored so many others. And most sociologists say that studies of the environment carried on by psychologists and medical researchers have been conceptually inadequate.

Elsewhere in this book you will have read arguments about genetic constitution and the environment of man conceived of as an organism. I will concentrate on the concept of environment as it is understood by sociologists and in doing so I will have to make a few generalizations about disciplines and the ways in which their practitioners operate which will only be broadly true. But this is one of the reasons why I said it is a complex area.

The rat in its maze

The arguments about the proportions of the variance in test scores attributable to heredity and to environment *all* contain some idea that the central nervous system *in interaction* with its environment leads to the development of a mind. Both are entirely indispensible; in this sense the only answer we are sure of is 100 per cent + 100 per cent. Nobody knows anything about the specific relationship between genes and qualities of mind. We have to make inferences about how identical gene-structures (identical twins) have coped with different (or similar) environment. So, at least until we know something about genes (for example, how many are there to the nearest ten thousand?) and their relation to cognitive activities, we are limited entirely by the quality of our thinking about the environment. I shall ignore the possibility that we can spoil our research by inadequate analysis of data, although this can also be crucial. For example, Farber was able to reject the Burt social-mobility findings purely on the grounds of inadequate statistical treatment. To put the crux of the argument; much of the psychological and medical research that has been carried on relating to this question has looked upon the environment as being physically external to the individual. For the sociologist (and social anthropologist, who will henceforward be included in this designation) the most important aspect of the environment is in the head of the individual. That is, it consists of the owner's views of the world around – his ideas about what surrounds him and how he may manipulate it. Instead of making use of the models of the environment available in sociology most 'after the event' research on the nature–nurture question has tended to use one that has done yeoman service in psychology – that of the maze. This is not to say that psychologists have not tried to use the *variables* suggested by sociologists. It is mostly that the method–ideology employed in 'operationalizing' (or measuring) them was inappropriate. One can imagine an experiment in which groups of rats were brought up in different mazes which contained 'tasks' of different levels of complexity. If we measured their intelligence at the end of a given period we might expect to find different average intelligences. Given some reason for assuming similar basic constitutions at the start we can explain the differences in average scores as a function of the different 'quality' of the maze experiences. That, crudely, is the model for data collection on identical twins and the model of the environment, also, for much of the other research on 'social factors'. But it is fatally flawed because people are not rats – they are not even vaguely like rats in the characteristics which matter.

The examples of how the environment has been misconceived in psychological research and discussion are endless, but the classic example is contained in an attack upon 'environmentalists' by a colleague of Sir Cyril Burt. In discussing the research work of Burt, Conway pointed out that from 1922 studies have revealed clear differences between the average intelligence scores in the different social classes which ranged in an order identical to the order of prestige of those social classes. Moreover this ranking of average scores is the same today as it was in 1922.

However, Conway states:

During the last thirty years, the environmental differences have greatly diminished. Both the economic and cultural conditions prevailing in the humbler classes have undergone vast improvements. If the environment was the chief source of the difference between one class and another, we would naturally expect that the IQ difference would likewise have diminished.

There are three important misconceptions in that quotation which may be summarized in the following ways:

1. *That the environment is external and concrete.* The only way in which the 'humbler classes' might be said to have 'narrowed the gap' is in terms of ownership of goods – things that are concrete and external to the individual. The allusion to increasing 'cultural' similarity presumably relates to 'high culture', posh mental activities or culture with a capital K; admittedly in the mind, but only a small part of the sociologists' definition of culture. It is probably not an important part except in so far as such pursuits have higher social status and a specially valued place in the school system.

2. *That environmental variables are discrete entities the distribution of which across the social status continuum has a linear association to a cognitive hierarchy.* This is more difficult to see in this particular quotation but is usually present in all writings that make maze-like assumptions about the environment. Here, it is present in the idea that if the social classes come closer together (i.e. the variables that comprise one become more like the variables in its neighbour) then the 'intelligence' the environment produces would become more similar. This is not necessarily so on the sociological model. For example, in one study, I found that within a narrow band on each side of the manual/white-collar social class division there was a tendency for home ownership to relate to 11+ success even though the owned homes tended to be of inferior standard to the rented council houses. The point here is that the concrete variable 'housing' was only an indicator of social views and intentions of parents. On a wider social class band, housing would continue to relate to 11+ success but it would also be possible to interpret it to mean quality of material home environment. Thus, the variable is not a 'thing' analogous to a particular kind of trapdoor problem for a rat, its relevance to the research question will be dependent upon the context in which it occurs.

3. *The use of a cognitive deficit model of cultural difference.* This assumption is usually less blatant than in this example but does

occur and does tend to obscure objective analysis. The problem here lies in the extent to which it is reasonable for us to assume that lower-class thinking is *inferior* rather than just different. Certainly, the educational system tends to 'assume' that this is so and, therefore, there may be some justification for us to act as technician to the system and accept its prejudices. Conversely, if we are interested in *changing* the system as opposed to maintaining it, then the technician approach will be unnecessarily blinkered. Sociology certainly offers no justification for assumptions that specific cultures are inferior to others. All of these conceptual errors can be seen to some degree in most research and particularly in anti-'environmental' writing. You might like to look for examples in the following attempt to find 'contradictions' in the environmentalist expectations of environmental consequences:

What do those who deny the importance of genetic determinants in the causation of individual differences in intelligence reply to the argument from regression? The answer to this is two-fold. Most environmentalists do not reply to it at all because they disregard this rather astonishing fact completely. Those who do pay some attention to it argue along the lines which may at first seem reasonable, but which can be seen to be contradictory. They say that the fact that fathers are highly intelligent and occupationally successful does not guarantee an optimum upbringing for their children; other things apart from money are important, such as parental interest in the child's progress, intellectual companionship, and other non-material factors. True, but on this argument high parental intelligence and above average material possessions would have to produce a very strong negative set of IQ determinants. For consider: parents having an IQ of 140 have on the average come from a social background materially and intellectually inferior to that they themselves provide for their children; this can be inferred from the fact that some 30 per cent of them would have been upwardly mobile socially. Yet their children have a mean IQ something like 20 points lower than their parents, although provided with all the advantages we are told in other contexts constitute the environmental determinants of IQ! If then the immaterial environmental determinants above are much more frequently found in dull and socially impoverished parents, and are so important that their absence in the affluent families produce a drop of 20 points of IQ in their children and their presumed presence in the dullest unskilled group produces a rise of 7 points of IQ in their children, why is it that middle-class children (and parents) are so significantly superior in IQ to working-class children (and parents)? This class difference is usually 'explained' in terms of precisely those material and intellectual advantages which characterize our higher professional parents. In other words, the environmentalist critics again want to have it both ways – factors are alleged to be important in relation to one set of facts, unimportant (or even negative in their effects) in relation to another. Such reasoning is clearly illogical: the facts are not compatible with any environmentalist hypothesis yet proposed, but are exactly as demanded by a theory combining hereditary with environmentalist determinants in the proportion of 4 to 1 (Eysenck, 1971).

Now, if the social environment really was like the maze in which the rat learns to accomplish 'tasks' and if there were a range of mazes which arranged themselves in order of difficulty analogous to an imagined order of 'quality of intellectual and motivational stimulation' in the social classes then this would be an excellent point. The sociological perspective leads us to a much more complicated view than that.

But before developing the point it is perhaps worthwhile to assert that there is no reason *at all* why the same 'factors' cannot have altogether different consequences depending upon, at the group level, the context within which we imagine it to operate, and at the individual level, the response which any specific individual chooses to make towards it. (Swift, 1967; Wiseman, 1968). This is because 'it' is not a thing at all but a construct in the observer's mind who had chosen to isolate it, by means of some 'measurement' technique, from the unlimited number of abstractions he could make from the world around.

This is not to repeat the point phenomenological sociologists have made about the positivistic and mechanistic views of social reality which behaviourist psychology tends to impose. It is to point out that not only is intelligence not a 'thing', neither are any of the 'factors' in the social environment that are likely to matter to the development of intelligence. And the model of research used by psychologists does tend to treat the environment as if it were.

In recognizing that these 'variables' are not discrete entities with special effects of their own, but concepts in our minds which we use to explain the workings of other people's minds and which are, therefore, only meaningful in their setting, we are faced with a serious methodological dilemma.

Two techniques of data analysis favoured by researchers using the maze model of the environment, are factor analysis and regression. Such strategies demand certain assumptions about the variables which the sociological model declares to be impossible. For example, the variables must be discrete and independent and distributed in the population according to the statistical curve of random error – in behaviourist psychology, cosily called the normal curve. Unfortunately, these are the two very assumptions which may not be made about them. A variable called 'parental support for school' will depend for its effect upon the child on what the parent understands by school. The lower working-class 'support' represented in the view 'If he is naughty you belt him and then I'll finish the job when he gets home' is 'supportive of school' but not quite the same as 'How do you advise me to approach a discussion of his homework problem' which is upper-middle-class 'support of school'. Similarly, the sociological perspective assumes that views of reality and values are structured; they hang together amongst groups of people rather than being randomly distributed. If we then manipulate our data in ways which are specifically denied by the ideas we have about what the data represents, we should not spend too much time worrying about 'inconsistencies' in our eventual data.

A sociological view of the environment

As good an introduction as any to the psychological theory of the development of self which is implicit in sociological thinking has been offered by Dubos, a micro-biologist:

> Since the real world . . . is perceived naturally through the senses, man's awareness of it can be studied scientifically by analysing the mechanisms through which the body registers environmental stimuli. As far as the perceiving organism is concerned, however, what really matters is its experience of the total environment, rather than the processes through which it apprehends reality.

Furthermore:

> Man ... does not react passively to physical and social stimuli. Wherever he functions, by choice or by accident, he selects a particular niche, modifies it, develops ways to avoid what he does not want to perceive, and emphasizes that which he wants to experience.

This is why we must say that the environment of the individual is in his head. How he 'behaves' depends upon three aspects of the environment in his mind – how he cognizes or understands the situation, how he values what it is in relation to how he will behave, and the kind of pleasure or pain he expects from it. All these elements are, to an unknown extent, a function of his previous learning. Finally, we must also make room for the intervention of a conscious process of decision making on his part.

It follows from this general view that what the tests aim to do – sample cognitive processes in the head – is only another way of sampling his experiences and the 'reality' that he is making of them. If it is true that 'genes do not determine the characteristics by which we know a person; they merely govern the responses to experiences from which the personality is built' (Dubos), then the process by which a self or mind develops is only explicable in terms of what that self has 'made' of its environment, i.e. of its previous human interactions.

I have said that the social environment is not external to the subject; it cannot be broken down into variables which are discrete in the statistical sense of being independent, or specifically causal regardless of context. Nor can we assume that these are distributed across the *status continuum* at levels of strength, efficacy, intensity or frequency which associate positively with position on that *status continuum*. And finally we cannot accept quality distinctions between cultures. Of these I find difficulty in justifying only the final characteristic and will deal with it separately – the others will self-evidently flow from the model of social process.

This theoretical point of view has been put most persuasively by linguists, but is now developing within American educational writings in response to the blinkered scientism of the Jensenist heresy. The linguistic argument is very straightforward. It rejects the concept of verbal deprivation as it has been implicitly applied in explanation of 'lower' intellectual achievement of ghetto children. Since the linguistic behaviour of ghetto children *is* the evidence (i.e. test scores) which Jensen uses to hint at the possibility that genetic inferiority of Negroes can be used as an explanation for failure of 'remedial' programmes for the verbally or culturally 'deprived', this can be a critical point of debate. On this argument it is simply the tests that are inferior for some people. Labov summarizes his view that linguistic specialists are united in rejecting Jensen's claim that the middle-class white population is superior to the working-class and Negro population in the distribution of what he calls conceptual intelligence in the following way:

> The knowledge of what people must do in order to learn language makes Jensen's theories seem more and more distinct from the realities of human behaviour. Like Bereiter and Engelman, Jensen is handicapped by his ignorance of the most basic facts about human language and the people who speak it.
>
> There is no reason to believe that any non-standard vernacular is in itself a handicap to learning. Our job as linguists is to remedy this ignorance. . . . Teachers are now being told to ignore the language of Negro children as unworthy of attention and useless for learning. They are being taught to hear every natural utterance of the child as evidence of his mental inferiority. As linguists we are unanimous in condemning this view as bad observation, bad theory and bad practice.

Tests of linguistic behaviour on which psychologists impose some assumptions about 'intelligent thinking' are the evidence for conclusions about average social class or racial inferiorities. If this imposition of assumptions is invalid or if we are making excessively crude assumptions about the equivalence of ghetto Negro with ghetto Mexican cultures, of Liverpool dockland with Hull fisherman or South Wales mining cultures we are being culturally parochial. That is, if there is much evidence for Labov's view that our tests still contain superficial language preferences of social class origin disguised as tests of thinking skills then the Jensen hypothesis is at best a possible final explanation to which we might fall back once we have cleared up the monumental inadequaces of our data. The efforts since the Second World War to 'clean up' intelligence tests by removing social-class specific *content* have really only scratched at the surface of this problem. Neither can we place much faith in research on motivation at the time of testing as opposed to during the process in which the habits of thought develop. Culture-fair tests are inevitably a chimera, if only because, as we saw in chapter 5, a test is a social situation. But this critique goes much further than establishing that. It implies that the validation procedures of psychometry (a good question is one which good pupils do well) and its close attachment to the social behaviour, which in school is often misinterpreted as intellective skills, can mislead us into a culture-bound theory of intellect that is only a reflection of the dominant behaviour patterns in society. If we accept that the language of culturally deprived children is not a basically non-logical mode of expressive behaviour we will find great difficulty in justifying the rules of language which we impose upon children in schools and tests designed to predict school behaviour. Although Bernstein has always insisted that his two linguistic 'codes' which have been loosely connected to lower- and middle-class culture cannot be ordered in terms of quality there is no doubt that educators and educational psychologists have tended to assume that the elaborated code was the appropriate mode for exhibiting higher cognitive functions and not just an elaborated *style*. If this has been an error we have reached a position where instead of arguing about how our data are to be interpreted we undoubtedly ought to be throwing it all away and beginning again. But before doing that we ought to wait until the psychologists have learned enough linguistics to come to grips with the problems which this rapidly developing science poses. From the sociological perspective it can only be said that the relativistic view contained in the proposition that the dialect of ghetto culture is a coherent whole, well adapted to the needs of its environment and carrying the capacity for all forms of cognitive behaviour, is entirely congruent with any theory of culture.

The cognitive style or problem-solving strategies of a particular child; the structure of his language; his view of, and approach to, formal learning, are specific instances of a culture. They are evidenced to us in the strategies of interaction which are played out in the classroom. As such we can only interpret them as intentional behaviour. By this I mean that the behaviour we observe (and this includes test writing) cannot be described in terms of the actions alone – one child's look of concern is another's dumb insolence – they must be interpreted in the light of the systems of meaning, firstly, of the culture within which the child principally roots his concept of self, and secondly, of the specific culture of the setting within which the interaction has been generated. That is, the rules of symbolic organization present in the cultures and intersecting in the behaviour of the child must be taken into account if we are to arrive at any propositions about how cognitive style develops in individuals. The task of course, is one for a psychologist and it is interesting to note how research on the development of cognitive style like that of Piaget or Bruner has become increasingly sensitive to the extent to which the findings on individuals are representations of culture. This may be compared with the grossness of the 'after-the-event' analysis of average scores underpinning the hereditarian research, which, because one cannot analyse culture after the event, has remained at an antediluvian stage of statistical analysis of hopelessly contaminated variables. In this kind of research one even finds correlational techniques employed to 'measure' the association between social status positions (see for example Eysenck, 1971, p. 67). This particular effort even went so far as to assume that there are only six occupationally based sub-cultures. If this were feasible, of course, there would be the beginnings of a case for the hereditarian use of psychometry. Then the 'control' of the environment through after-the-event statistical manipulation would become a relatively permissable device instead of the complete abdication of scientific method it really is. When we are using societal-level analysis of social structure it becomes essential to describe that environment in terms that are specific to the individual. At least, in so far as the main aspects are concerned. In this case, we would need to describe each individual's 'profile' on a range of cultural attributes – his orientation towards time, authority, human nature and tasks, the nature of his language, the strength of his motivation to achieve in general, and the aspects of behaviour that he defined as being worthwhile achievement. Of course, when we have done all that we will have an incomparably better measure of his exhibited ability than any test so far devised. And that, in a sense, is the point of the argument. Intelligence tests have some value as predictors of that social behaviour known as adaptation to education: as such they are analytically useful reifications of social process. Unfortunately, their results may also encourage us to make incredible biological assumptions of the form – people who are not in the habit of making use of classifications like 'animals', 'furniture', clothing' or 'foods' in conventional tests are not capable of conceptual thinking because of the genetically produced absence of adequate neural structures. If we do we must be prepared for linguists to tell us that there is no language or dialect that does not contain much more difficult conceptual tasks, that anyone who speaks one must inevitably possess greater capacity than that; and for sociologists to point out that the best explanation for our findings lies in the tester's inability to understand the culture from which he has isolated an aspect and upon which he has imposed his own culturally specific point of view.

Some commonsense conclusions on the relevance of circumstantial evidence to situations in which we are attempting to change the circumstances

The assumptions of conventional testing have been summarized in the following way: Intelligence A (IA) represents innate potential; as such it can be thought of as constituting an individual's intellective 'ceiling'. It is assumed that this ceiling is ultimately determined by the genes. Note that the assumption about genes is not couched in terms of the obvious point that they are a *necessary minimum.* Intelligence B (IB) represents intellective ability as exhibited in everyday life. Intelligence C (IC) is the sample of IB activities that are contained in an intelligence test.

IA sets the ceiling and experience prevents the individual from achieving it – we may call this 'the gap'. While this is the assumption about the relationship between IA and IB nobody ever looks at the data in this way – the argument is always about what contribution social experience has made to IC. It should, on the theory, be about how large a gap between IC and IA it has produced. But this would expose the scientism in the whole edifice of thought since IA, by definition, is unquantifiable. This is perhaps a sophisticated definition of scientism which I would normally define as an implicit view that 'if you can't measure it, it isn't there' – in this case it would be that 'if you can't measure it you assume that measuring what you can measure covers the whole question'.

Use of IC in discussion of questions about genetic endowment of social groups must assume that tests contain an adequate sample of IB. The sociologist and linguist are likely to ask, whose IB? The answer all are agreed upon is 'the IB that is rewarded in the educational system'. All are probably agreed that this has only a tenuous connection with behaviour 'outside' the school. Opinions differ about how tenuous the connection is. Certainly, intelligence and verbal reasoning test scores group within bands for a range of occupations in an order that clearly relates to the extent to which school-type behaviour is expected of the holder of the occupation. The differences of interpretation arise in two ways. Firstly, we have an argument about whether or not there is some 'ideal' structure of thinking skills that may be used to impose a measure upon observed problem-solving behaviour, like, for instance, Spearman's definition – the ability to educe relations and correlates. The sociologist would only be able to offer a discussion similar to that used by Eysenck (1971, p. 78) which relativizes the skills to the culture within which they are appropriate. Argument would centre on the boundaries of the culture. For example, most psychologists would agree that the use of the same verbal reasoning test in comparing aboriginal Australian children with upper-middle-class English would be inadmissable. The disagreements arise when a sociologist argues that the differences between upper-middle-class English and lower-working-class Liverpool English cultures may be just as great from the point of view of valued intellective skills, though more difficult to detect because of their superficial similarities. This latter proposition would be illogical to a strongly positivistic psychologist who has a highly mechanistic and deterministic view of the environment, the elements of which he was inadvertently assuming to be 'things' rather than aspects of the individual's 'self'.

Secondly, we may debate about the extent to which this ranking of occupational group scores is a consequence of the sorting function of the educational system as opposed to a consequence of the actual needs of the occupation. A partial alternative explanation to both lies in the validation procedure of test construction. We would expect that the hierarchy of respect towards thinking activities and verbal style which a culture evinces in its school system would roughly coincide with that which is implicit in its occupational status hierarchy.

The gap

'Hereditarian' research assumes I C to be an indicator of I A. That is, in addition to assuming that I B adequately samples I C it assumes 'the gap' between I A and I B to be similar in different people – presumably because genes govern the characteristics of mind. Research in which the environment is inadequately conceptualized finds only a small amount of variation in scores attributable to the environment. Consequently, possibilities for increase in I C are seen to be limited (and also prevented by test construction precedures).

An alternative approach assumes that genes predispose the individual to seek or respond to stimuli the experiencing of which leads to development of habits of thought. These are *socially* defined as 'intelligence'. Our problem then becomes one of identifying the culturally specific definitions of I B.

Use of circumstantial evidence

In any research problem we should ask why we are asking the question and why we are asking it in this way. The answer to the first question is that our egalitarian society is concerned to improve the opportunity it offers all individuals to develop their capabilities. The major efforts made in the USA and the minor ones in Britain to compensate for the more glaringly inadequate opportunities follow from this, are expensive and have been fairly spectacular failures. The only possible reason for asking the research question in the form 'what *individual level* reasons can be found for *group* differences in thinking habits?' is that the answer may give a reason for these failures. Analysis of average scores of 'groups' of subjects which are either a figment of the observer's imagination or his attempt to isolate social categories, could lead to 'evidence' which may be interpreted as representing some deeper state of affairs in individual persons which caused the failures. It is important to recognize that there is a much better way of conducting research on the development of intelligence. We might study individuals to see how it develops and what factors of 'stimulation' have what kinds of consequence for it. This would provide us with evidence that is much less circumstantial and 'after-the-event'. It would be much less suspect from the point of view of its subservience to 'ghost-in-the-machine' explanations of past and present social process.

Nevertheless, we are entitled to look at what we have got and to make of it what we can. For the last decade there has been a general agreement to ignore the body of research using correlational analysis of test scores that was mostly carried out before the Second World War as being too inconsequential and scientifically inadequate to offer even vaguely reliable information on matters that have some value in educational decision making.

In the last two years the 'Jensenist heresy' has arisen, presumably because there is a need for a scientific explanation (however tentative) for the failure of counter-deprivation programmes. Failure in this case, must be defined as failure to improve the position, rather than failure to achieve some goal which may have been set too high. (As a matter of fact, some of the programmes have not been failures in this sense – but let us ignore that for the sake of discussion.) In such a situation we have two kinds of explanation – either the material (i.e. children) were inadequate, or the methods (i.e. teacher's knowledge of the development of intelligence and the means by which it can be stimulated) were inadequate. The Jensenist heresy with its belief in 75–80 per cent heritability suggests that total failures can be explained best along the former lines. But I find this difficult to follow – given the 20–25 per cent variation in test scores attributable to experience, then *total* failure ('... these special programmes had produced no significant improvement in the measured intelligence or scholastic performance of the disadvantaged children ...') (Jensen, quoted in Eysenck, 1971, p. 21) is *proof* of the inadequacy of the programmes as teaching method. Here is something a Jensenist can be sure of. Compare this with '. . . all we are left with are various lines of evidence, no one of which is definitive alone, but which, viewed together, make it a not unreasonable hypothesis that genetic factors are strongly implicated in the average Negro–white intelligence difference' (Jensen, quoted in Eysenck, 1971, p. 30). There is no claim of proof here – only a beautifully illogical piece of reasoning. A range of inadequate findings, because they point in the same direction can be added together to prove, not that there must be a similar fallacy underlying each of them, but that the tendency might very well be stronger than any single one shows. Even granted this summation of weaknesses, why prefer the weak finding – 'not unreasonable hypothesis' – to the incontrovertable one that *can* be drawn?

Accidental socialization

We know so little about how the mind develops and how it may be influenced through manipulation of its experience; computer simulation is making such inroads upon sacrosanct areas of human intelligence and exposing to us a recognition that much of what we think of as intelligence is no more than simple memorizing of previously accomplished tasks; we know so little about variations between cultures and of the relevance of culturally preferred verbal styles to 'necessary' thinking habits and hence, abilities, that we ought to call all previous socialization *accidental* as far as the development of cognitive style is concerned. For example, all the data summarized in the twin studies are of this kind. Thus, we could accept that over the three decades to 1960 accidental socialization has produced figures of 75–80 per cent heritability of test scores and assume that in the same three decades of the nineteenth century it was 90–95 per cent. We might then go on to assume that as the rate at which the education system affects the remainder of society increases, as our research on individuals teaches us more about how mind develops through experience, as we discover how to influence the chemical or psychological consequences of the gene areas and how to bring more of them into play, there will be an accelerating tendency for the average heritability element to decline.

If we do this we are faced with a dilemma in the use of

circumstantial evidence. We begin to recognize it for what it is – a description of the social *status quo*. Clearly it may be reasonable to use it to tell us where we stand, but how can we use it to tell us how we might change? We cannot; for that we need real evidence of the kind reported by Heber and summarized by Eysenck in the following way:

> But with all these qualifications the fact still remains that this experiment holds out the promise that by transforming the environment in a manner going beyond anything attempted as yet we may be able to raise IQs generally to a degree which would make it possible for the average person to benefit from a university education which under present conditions would be well beyond his ability (Eysenck, 1971, p. 134).

This research showed it to be possible that we might learn how to raise the intelligence of children by, say one standard deviation in a relatively short space of time. An IQ of 100 becomes IQ of 115; a child moves from a score that puts him on the fiftieth percentile to one which (prior to this intervention in the societal process of socialization) was on the eighty-fifth percentile – a secondary-modern B-streamer goes to university.

The problems of education are problems of what might be, of what can be attained as well as what ought to be sought. Circumstantial evidence tells us only what was – it cannot take account of the new knowledge that we are seeking, the new processes and structures of education we are developing, the changes in society for which we are preparing. It may satisfy one who wishes to maintain the *status quo*. It may very well subvert the intentions of the reformer, because it does not consider what *might be* – and education is *about* what might be.

Further reading

B. Bernstein, 'A public language: some sociological implications of a linguistic form', *British Journal of Sociology*, vol. 10, pp. 311–26, 1959.

J. Conway, 'Class differences in intelligence: II', *British Journal of Statistical Sociology*, vol. 12, pp. 5–14, 1959.

R. Dubos, *So Human an Animal*, Rupert Hart-Davis, 1970.

H. J. Eysenck, *Race, Intelligence and Education*, Temple Smith, 1971.

B. Farber, 'Social Class and intelligence', *Social Forces*, vol. 44, pp. 215–25, 1965.

E. Fraser, *The Home Environment and the School*, University of London Press, 1959.

W. Labov, 'Logic of non-standard English', in F. Williams (ed.), *Language and Poverty*, Markham Publishing Company, 1970.

D. F. Swift, 'Family environment and 11+ success: some basic predictors', *British Journal of Educational Psychology*, vol. 37, pp. 10–21, 1967.

S. Wiseman, 'The effect of restriction of range upon correlation coefficients', *British Journal of Educational Psychology*, vol. 37, pp. 248–52, 1968.

9 Nature versus Nurture: The Significance of the Revived Debate

John Rex

John Rex was born in South Africa in 1925. He was deemed undesirable as an inhabitant of or visitor to Southern Rhodesia in 1969. A Lecturer at Leeds and Birmingham Universities, he became Professor of Sociology at Durham University in 1964, and at Warwick University from 1971. He is the author of a number of books on the sociology of race relations.

The misuse of quantitative methods

Quantitative methods in the social sciences have much to commend them. If used sensitively and with understanding, they ensure that the observations of any one scientist or observer may be replicated by another. The great danger, however, is that if they are used insensitively the social scientist may seek to quantify for the sake of quantification, and, if the issue under discussion is not capable of easy quantification, it is likely to be put on one side and replaced by another. Too often, quantitative social scientists give us exact but irrelevant answers to the questions we are asking.

When we speak of 'sensitivity' on the part of a social scientist we refer to his awareness of the relationship of that which he measures to a body of theory and, through this body of theory, to other measurable concepts. We also refer to the sensitivity of the social scientist to the meaning of human action for the participant actors whom he observes. In a word, it is a requirement of sensitive social science that the social scientist should be aware of the fundamental epistemological problem of the human studies, namely that, while natural science is an activity in which scientists have concepts about things, in the human sciences the scientist has concepts about things which themselves have concepts.

These observations have led in recent times to a revolution in the social sciences. It has been argued, for instance in relation to criminal statistics, that official statistics have not recorded the quantitative occurence of an act of a certain kind, but rather the numbers of those acts which other people such as the officers of the law categorize in a certain way. Similarly, most demographic and ecological statistics, when they are probed, turn out to refer, not to simple attributes of persons and their behaviour butrather to the way in which people are classified in practical situations.

The ideology of empiricism and operationalism seeks to avoid these problems by arguing that, since there are no 'essences' in the world to be measured, the measurable variables are definable simply in terms of the tests which are to be used to make a measurement. Thus, for example, a foot is that which is measured by a foot rule, intelligence is that which is measured by intelligence tests, and so on.

What is called psychometrics is perhaps the least sensitive and the brashest of the empirical human studies, as may be seen from some of the preceding chapters. On the matters just discussed it seeks to get the best of both worlds. It claims, on the one hand, that it makes no assertions about essential intelligence and that what it refers to is simply measured intelligence. On the other hand, however, it pretends that this measured intelligence has no reference to practical, social and political contexts.

The first point which we need to make here in asserting the need for a genuine sociology and psychology which is theoretically founded and aware that it is dealing with meaningful action, is that the position of the psychometrist, who pretends that what he says about measurable difference has no practical significance, but simply refers to facts which may be classified as true or false, is untenable. The intelligence tests to which he refers are used in practical contexts, as a matter of empirical fact to assign children to different forms of education, to choose between one individual and another in job placement, and generally to set one man above another. It is therefore not possible for a psychometrician to say 'I am merely facing up to a scientific truth, albeit an uncomfortable one'. What he does when he rates individuals or groups of individuals on a scale of measured intelligence is to say and to predict that one group of individuals rather than another should have privileges. It is of little use, therefore, that a writer like Eysenck should protest that there is a total disjunction between his scientific observations and his moral views. Scientific observations have political implications and the scientist should beware that that which he reveals may contribute to, or ease, human suffering. This, of course, is not in itself an argument for not facing up to facts. It is, however, an argument for the human sciences to beware of jumping to rash conclusions on the basis of simplistic scales of measurement.

If we look at the popularizations of Jensen's ideas on racial differences, we find no such circumspection. The problem not merely of the nature of intelligence, but of its causes and correlates, is over-simplified to an almost incredible degree. According to Eysenck, for example, the issue is between 'interactionists' like himself who believe that both nature and nurture contribute to test intelligence and those whom he calls environmentalists who are supposed to hold that intellectual capacity is solely the product of a few easily measurable environmental factors, such as amount of income, type of residential neighbourhood and years of schooling. All that is necessary to disprove the 'environmental hypothesis' and thence by implication to prove its opposite is to show that when these few environmental factors are held constant, observed differences between individuals or between groups, are maintained.

The reply to this rests first of all upon the recognition that intellectual arguments which go on between social scientists on the matter of nature and nurture do not involve one side which is simply environmentalist in Eysenck's usage of that term. Those whom he specifically attacks, namely, UNESCO and a distinguished line of social scientists who have worked since 1945 to expose fallacies of racism, have always recognized that there is ground for supposing that there is a genetic component in measured intelligence. Moreover, they have also recognized that the processes of selection and isolation do lead to groups of men having different gene pools. What they have disputed is that these differences are so great that manipulation of the environment is not capable of fundamentally altering them.

More important than this, however, is the naive belief that

environment can be reduced to an index based upon the few quantitive variables (some of the failures in this belief are explored in chapter 8). Such a view naturally commends itself to those taking a simple mechanistic model, who refuse to accept that the relationship between the performance of acts and the events which precede and follow them may be meaningful rather than simply casual. But even if we take the assertions of psychometric empiricism in their own terms, it seems clear that the argument that any measured differences not assignable to the size of income, type of residence or length of schooling, must, due to genetic factor, shows a remarkable over-eagerness to jump to conclusions.

When such a method is applied to the comparison of Negro and white intelligence in the United States, there are obviously a great many other variables which should be controlled. They cannot all be summed up under something as simple as 'motivation' (one of the factors discussed by Jensen and Eysenck). It matters as is shown in chapter 3, that the Negro group is continually exposed to a picture of American society, in which, if it is not subject to racial exploitation, is, at least the object of benevolent paternalism. Moreover, if one looks at the content of schooling, it is clear that Negroes studying American history in which the heroes are all white, are bound to respond differently to their education from their white school-fellows.

Curiously, Eysenck asks that those who oppose him should offer 'experimental' evidence. Fortunately, neither he nor his opponents are able to undertake such experiments. Since, however, the crucial variable is the difference between white and Negro history and the fact that Negro history involves the fact of slavery, experiment would mean subjecting the group of Negroes to white experience over several hundred years, or subjecting a group of whites to Negro experience. The empirical study which holds constant, size of income, type of neighbourhood and length of schooling in the United States of the present day, therefore, should in theory be supplemented for an experiment in which the peoples of Africa conquer, capture and enslave some millions of European and American whites under conditions in which a very large proportion of the white population dies and in which the white culture is systematically destroyed, and in which finally a group of emancipated whites living in 'good neighbourhoods' are then compared to their Negro masters. It is not sufficient to brush aside this assertion, merely by saying that we should not draw conclusions from 'hypothetical experiments'. The fact is that the differences in the history of Negroes and whites are a factor of immense significance and that any statistical reasoning which leaves them out can reach no conclusions of any value whatever.

One of the difficulties which the empiricist has, of course, is that he deals only in the external attributes of individuals. He cannot concede that intelligence or any other attribute of the individual may be understandable in terms of the meaning of an individual's environment to him. What has happened to the Negro over several hundred years is a process of Sambofication, a process by which he is first stripped of all identity and then forced to become the happy, shambling and incompetent child of the slave owner. White society maintains this stereotype of the black, and the black behaves in accordance with its expectations. As Elkins, amongst others, has pointed out, the nearest comparable process, of which we had to have evidence, is what happened to the inmates of German concentration camps. Similarly, the environment of a young, educated American Negro today is apprehended as one in which he seeks politically for a new identity and in which white education and the white police are equally agencies which undermine that identity.

This is only one aspect of the meaningful environment of minority groups in the United States. What is necessary before we can draw conclusions about the performance of one group or another is that we should understand on a meaningful level, the type of relationship which a minority group has to American society. Thus, for example, one cannot regard the relationship of Negroes, the descendants of slaves, to that society as qualitatively the same as that of American Indians, sidetracked from absorption by their life on reservations, or Oriental immigrants, who might be descended from indentured labourers, and who are very often engaged in one or other minor commercial occupation. Any acquaintance at all with the literature on the sociology of plural societies makes it clear that, given the very different relationships to a social system which minorities might have, it is quite meaningless to compare them as though the only environmental differences between them are those of socio-economic status.

The insensitivity of the psychometricians fails to take any account of these complexities of the real world. Their blind use of IQ test data, which on the surface resembles a valid attempt at quantification, leads to conclusions which are not only erroneous but lack any kind of reality. Unfortunately these conclusions cannot be dismissed as irrelevant but, as we shall see below, they have profound social consequences.

The rebirth of racism

It should of course, be noticed that Jensen and his supporters do not claim that they have reached final conclusions. At best they say that it is important that certain hypotheses should still be regarded as open to test. But, the mere assertion that these hypotheses are important in a scientific sense is taken by many to mean that the notion that there are genetically based differences in intelligence between the races is no longer simply a notion of racists. A view of racial inequality is then revived which was common in the early 1930s, but which was discredited in democratic countries after the defeat of Hitler. The main problem which we have to face then is one in the sociology of knowledge as it effects race relations. What difference does it make to our total political situation that scientists appear to be unconvinced that racial differences in intelligence are not innate.

In one sense it may be argued that this fact is not particularly important. After all, there have been long periods of history in which nations and other groups have exploited each other without seeking any scientific validation for their views. Moreover, it is true, particularly since 1945, that very few politicians indeed have claimed that discrimination is justifiable because of inherited differences in intelligence. Nonetheless, what we shall argue here is that, given our culture and the set of beliefs in that culture about the nature of science, the basing of racial inequality on a scientific proof of differences in intelligence makes that inequality far more permanent and durable than it otherwise would be.

At the lowest level, individuals find themselves in competition or opposed to members of ethnic groups and come to express hostile sentiments towards other groups. Action in these matters usually precedes that of rationalization. It is only as the individual seeks validation and justification for his views amongst his fellows that the process of rationalization begins. Very often the highest level of rationalization which is reached is the sharing of an anecdote about an out-group amongst members of an in-group. Yet the process whereby such low-level rationalizations are achieved connects with rationalizations at a higher level which are provided by the definitions of social reality contained in media messages and in the statements of influential local leaders, such as ministers of religion and political leaders.

All individuals finding themselves in new situations seek to arrive at a shared definition of these situations with their fellows. Most of us do not really feel that we know what the world, and particularly the social world, is like until we have corroborated our views by checking them with those of our intimates, but more systematic rationalizations occur when individuals are affected in their definitions of reality by what the newspapers or television programmes have to say. In a quite different and more trivial field than that of race relations, for instance, it is interesting to notice how the judgement of the followers of the main spectator sports are limited and shaped by the picture of the sporting world which is provided by sports journalists.

The level at which the popular press, radio and television provide rationalizations for action is of course itself fairly unsystematic. In the field of race relations, one does not expect to find systematic argument about the nature versus nurture controversy in the columns of, say, the *News of the World*. What one does expect to find there are anecdotes; yet such anecdotes may be of great importance. To give an example, on the Sunday evening which followed the first of Mr Enoch Powell's speeches on the subject of race, BBC television presented a programme in a story series about a local councillor, in which an outbreak of stomach trouble was traced to an Indian restauranteur who offered his customers cat food instead of fresh meat. Such a programme may be expected to have had the effect of reinforcing one of the best known myths about Asian immigrant communities. The belief referred to here is widely held, but its inclusion in a serious BBC programme gives it a kind of legitimacy which it could not otherwise hope to have.

Those who control the popular mass media may or may not be aware of the extent to which they help to define social reality for their public. There may, on the one hand, be cases of sheer manipulation of public opinion, but on the other, there will be cases in which the media merely reflect what are known to be popularly held beliefs. In either case, however, popular mass media do have some sociological significance.

Just as the images of reality which result from face-to-face sharing of experiences are influenced and reinforced by the messages of the popular media, so these latter are affected by more systematic formulations to be found in other 'more serious' publications. What is said in a loose language of sentiment in the popular papers and programmes is said more systematically and intellectually at a higher level in the quality papers and quality television and radio programmes. Such programmes attempt to deal with problems of current affairs in terms of relatively consistent and coherent sets of beliefs, even though the set of positions which they occupy falls far short of a systematic scientific statement.

On the whole, the quality media have not been entirely useful in promoting racism. They represent the public face of a culture and society and must adhere to a minimal standard of political beliefs from which those which foster racialism tend to be excluded. Such a situation, however, might still change if those occupying positions of authority and moral respect by-pass the quality newspapers and other media and lend their support directly to reinforce the racist definitions of social reality which are to be found at the grass roots.

One of the features of most advanced industrial societies is the down-to-earth outspoken politician who is willing to say things which are believed, but which have become taboo in quality discussion. We sometimes say that such politicians produce a gut reaction, rather than an intellectual response, but for all that, their political speeches are of the greatest importance in that they raise the level of respect which is to be accorded to racist myths and anecdotes. Neither the quality papers nor the populist politicians' utterances however, have anything like the systematic deterministic nature of scientific statements. Indeed they do not form any kind of coherent theory at all. It is only when they in turn are subject to rationalization that the higher forms of systematic knowledge become related to political action. At this level we find political intellectuals who write books and argue in public places, defending their beliefs in a systematic way. They may argue from a cultural, historical, sociological or religious standpoint, or they may argue from the point of view of science.

While the big taboo on scientific theories of race was maintained 1945–67, individuals seeking to rationalize racial discrimination would refer to the differences in history and culture between groups or occasionally to religious belief systems. In most cases the picture of the world which they were able to draw was not a deterministic one. Man may after all change his culture or his religion even though there have been some cases, as in South Africa, where doctrines like that of predestination have been used to suggest that there are immutable differences between groups which are ordained by God. While, however, religious beliefs usually involve some possibility of the individual's transformation through a process of salvation, it is a feature of scientific theories about human nature that they tend to be deterministic. Increasingly, in our culture, we are encouraged to believe that what we can do depends upon scientific possibilities and those scientific possibilities are held to be determinate. Thus, when it is said Negroes are genetically inferior to whites, the differences which exist between them are thought to be entirely immutable. Undoubtedly it is the consciousness of many scientists that this would be the effect of their talking loosely about racial differences which has led them to be very careful and cautious in what they say.

If, then there is nothing like a conclusive case for the genetic determination of racial differences in intelligence, as shown in chapter 5, why is it that at this moment in our history, theories which have remained dormant for thirty years, are suddenly revived? The answer to this must surely be that the enunciation of such theories fulfils a political and social need. It may be suggested that it is precisely when there is a gap between theory and practice in matters of race relations that the support of science is sought in order to bridge this gap. Thus, in the United States, the arguments of Jensen have been used against the acceptance of various poverty programmes. The popularization of his ideas by C. P. Snow and Eysenck may also serve to provide rationalizations for racial inequality in contemporary Britain.

Interestingly enough, these are not the only scientific theories which help to validate racialism and inequality. One significant phenomenon in our scientific history has been the revival of 'biological' explanations of criminality (e.g. the 2y chromosome theory). Another is the body of ideas associated with the words, ecology, conservation, pollution, etc. In this latter case human political judgement is no longer considered, even as an intervening variable. The ills of the world are simply explained as being due to inexorable scientific laws.

The crux of the sociological argument about racist biological theories therefore is this. On the political level societies may pass through periods in which there is no great need for any kind of theory which emphasizes the differences and incompatibilities between different ethnic or religious or 'racial' groups. In such periods popular maxims and anecdotes will affirm the essential similarities between men, and informed opinion will deplore the political behaviour of the small minority of disturbed persons at home who are 'prejudiced' and of governments abroad which work on a basis of racial supremacy. As strains in such a society develop however there is a groundswell of opinion in which popular maxims and anecdotes guiltily and uneasily spotlight racial differences. Thus in Britain, for example, towards the end of the 1950s racist jokes began to be heard in working men's clubs.

This first guilty snigger of racism however gained a new significance when leading politicians of both parties began to include hostile references to immigrant minorities in their speeches. Some like the late Lord Carron included ambiguous references to immigrants in the course of general attacks on government policy. Others like labour leader Robert Mellish found themselves drawing attention to the contrast between political ideals and the realities of the situation with which they were forced to deal at the level of local government.

In the next stage, however, the diagnosis of the problem became more systematic. Mr Enoch Powell, widely recognized as an intellectual in British politics, argued that basic issues were becoming suppressed and set out his own arguments as to the way in which the presence of a large proportion of immigrants in the population of British cities must undermine British culture.

Although Mr Powell's speeches produced widespread public support, they neither claimed that immigrants were in any sense inferior or that the differences between them and their British hosts were innate. The effect of world scientific opinion and the work of UNESCO since 1945 was such that responsible politicians were wary of committing themselves to such views. Mr Powell therefore based his case on the inadequacies of official statistics which, he argued, underestimated the number of immigrants, and the clear cultural differences between immigrants, particularly Asian immigrants, and a pseudo-sociological concept of territoriality.

The possibility always existed, however, that the immigrant population of Britain would become something of an underclass, deprived of housing and employment. Mr Powell's speeches gave only marginal help in fostering this situation. True, his insistance on the impossibility of assimilation gave some kind of basis to those who wished to discriminate. But the argument was never as water-tight as it could have been had it been based upon a theory of the biological basis of racial difference.

Whether they have intended it or not the popularizers of Jensen in Great Britain have now fundamentally altered the situation. The politicians who favour discrimination may in future argue not merely that Indians, Pakistanis and West Indians should not be assimilated and given equality of opportunity. They will now point to Eysenck's or Jensen's work and argue that assimilation is impossible and that equality of opportunity can only guarantee that Negroes at least will under such circumstances find their own level. There will of course be little hesitation about invalidly applying results from tests on American Negroes to *all* blacks in Britain. But in all likelihood it will not be long before American experiments are replicated with Indian, Pakistani and West Indian subjects in Britain.

The account of the structure of a society's belief systems which we have given here is of course speculative. It is a complex hypothesis about the way in which ideas are related to action and social structure in advanced industrial societies. The validation of such an hypothesis would have to take place through consultation with the historical record. To the psychometrician whose approach to this problem would be based on the measurement of prejudiced attitudes and the correlation of these with other measurable attributes, such an hypothesis is too complex to be capable of testing. Unfortunately historical reality is complex and its truth is unlikely to be discovered by the techniques and methodology of dogmatic scientism. The function of such scientism is not to discover the truth but to prove the inevitability of the *status quo*. The psychometricians, of course, would dismiss any such notion and would claim that they were merely describing their objectively gathered facts. However, the nature of their 'objectivity' – the values inherent in the concept of IQ, the simplistic notion of the environment does not bear close examination as several of the contributors to this book have demonstrated. What is argued here is that the misrepresentations of the psychometricians is not simply a matter of a random 'mistake' but is directly related to the beliefs of the society in which they operate.

Further reading

S. M. Elkins, *Slavery: A Problem in American Institutional and Intellectual Life*, University of Chicago Press, 1968.

D. Hiro, *Black British/White British*, Eyre & Spottiswoode, 1971.

L. Kuper and M. G. Smith, *Pluralism in Africa*, University of California Press, 1969.

J. Rex, *Race Relations in Sociological Theory*, Weidenfeld & Nicolson, 1970. This is a comprehensive introduction to the sociology of race.

M. G. Smith, *The Plural Society in the British West Indies*, University of California Press, 1965.

P. L. van den Berghe, *Race and Racism*, Wiley, 1967.

P. L. van der Berghe, *Race and Ethnicity*, Basic Books, 1967.

Conclusions: Intelligence and Society

Martin Richards, Ken Richardson and David Spears

Martin Richards is University Lecturer in Social Psychology at Cambridge. Though originally trained as a zoologist, he now investigates the development of children in the early years of life in the Unit for Research on Medical Applications of Psychology.

Ken Richardson originally trained as a teacher and later graduated in biology and education. He now does research at The Open University on the effects of environmental change on brain biochemistry and on the neurochemical correlates of learning.

David Spears took a degree in biology, physiology and psychology in London and worked in experimental psychology before taking up his present work on the metabolism of the central nervous system at The Open University.

Some general themes

The preceding chapters of this book have discussed many aspects of the problems of race, social class and intelligence. Here we will not attempt to provide a summary, indeed it is quite clear that the debate does not lead to simple conclusions, but rather we shall try to draw out some important themes from the discussion and then look more generally at education and society.

Much of the current confusion in the discussion of intelligence seems to arise from a dominant tradition in psychology which has tried to analyse people and their behaviour on a model derived from classical physics and chemistry. This tradition has attempted to find objective and static measures of highly complex and dynamic entities. All too often the gap between the 'objective' measure and the concept it stands for is so wide that the whole of the ensuing debate becomes highly misleading. This process is clearly evident in the attempt to reduce the notion of intelligence to the result of an IQ test. Here the whole of cognition, language, thought, and conceptualization is reduced to a single figure which expresses an individual's standing relative to his peers. Professor Swift has discussed the same phenomenon in the use of the concept of environment and has demonstrated the distortion involved in trying to reduce a person's perception of the world and the world's perception of him to a single variable which is then treated as if it were a physical constant. Again, in chapter 6 we saw how misleading it is to consider the development of intelligence in the framework provided by classical Mendelian genetics. It may be true that a scientist learns little about a phenomenon unless he can measure or count it but it is equally clear that there are many things which can be measured and counted without teaching us anything. As John Rex has emphasized, we need some sensitivity and understanding as well as an ability to do arithmetic.

Another feature of the traditional approach to the social sciences is a characteristically narrow and arbitrary definition of what constitutes a problem and the kinds of evidence which are held to be relevant to its consideration. Many of the contributors to this book have gone well beyond the traditional confines of the race–IQ debate and found much that is highly pertinent but which has previously been disregarded. So while the Jensen–Eysenck–Burt–Shuey school sees IQ testing as the objective measurement of intelligence, we have also considered it as a social situation where a member of a minority group acts in accordance with his views of the majority's evaluation of his group's worth. As Peter Watson has pointed out, this kind of evidence can lead to a very different interpretation of the lower scores that some black children achieve. Even more striking is the psychometricians' blindness for the social context of their work – a point we will discuss in more detail later.

This limitation of the framework for the discussion of intelligence does not rest on any obvious or particularly objective criteria – it seems more a case of scientific tradition – and therefore further undermines the claims for objectivity so frequently made by the psychometricians.

Another point that stands out from the earlier chapters is the extent of our ignorance of the nature of intelligence. Within traditional psychology, IQ has come to stand for intelligence, and its estimation and correlation with social and biological variables have taken all the attention and have excluded work on the nature and development of intelligence. So we find ourselves in a curious position. Though we know a great deal about the distribution of IQ scores in the population, the extent of variation in scores and their relationship to other variables and the statistical structure of test performance, we have very little idea about the kind of thing the tests are alleged to measure and how any individual acquires it.

If we wish to have a realistic debate about individual differences in intellectual performance, we need two things – a description of intelligence and an adequate theory of its mode of development. Psychometrics is unable to investigate the nature of intelligence because it is concerned with the relative ranking of people on criteria derived from the values of a selective educational system. For this tradition intelligence is what the tests test and high intelligence is doing well at school. We cannot use this notion to examine the relationship between intellect and education because it presupposes a particular connection between the two. It also cannot deal with cognitive diversity in a way which approaches reality because it uses one or a very small number of dimensions of evaluation. Psychometricians might object and say that this disregards the contribution of factor analysis (a statistical technique which separates out a number of different factors involved in test performance). However, this is irrelevant to the present argument because these factors are properties of the test-taking performance of groups of children and not of individual children.

As any teacher or cognitive psychologist is well aware there is an immense diversity of cognitive style in children and adults. Different people choose different ways of solving particular problems and the diversity displayed far outruns the psychometricians'. A child's style cannot, of course, be indicated by the single number of his IQ score but it also fails to be adequately represented by a complete battery of psychometric tests.

Varying cognitive styles are useful for different things. High-IQ people tend to be good at straightforward logical problems that have one correct answer. They do not do well when dealing with uncertainty, when probabilistic strategies are needed, and they often do even worse when a series of quick judgements of this kind are called for in a brief period. This is the busy executive at his desk with two phones ringing and his secretary at the door demanding an answer. Cognitive styles often associated with high IQ scores seem to reduce people to complete inaction in situations like this.

What is required for a realistic debate is an abstract theoretical description of intelligence in terms of rules and knowledge which is independent from the values of our educational system. As mentioned in chapter 1, Piaget's work gives us a starting point for just this enterprise. Unfortunately, we do not have space to elaborate on this research here, instead we must refer the reader directly to Piaget's work, but we should note that he has established a very healthy tradition of work on cognitive development. He provides a description of intelligence from earliest infancy to adulthood rooted in a development theory which is at least consistent with current views of biological development. As we saw in chapter 6, this is something which the psychometric–heritability work fails to do. It cannot cope with the known complexities of development because it rests on the false typology of genes and environment. Such a dichotomy is quite foreign to Piaget's idea of development. The important point here is not the details of Piaget's theories, indeed it is clear that some of these are wrong, but the kind of theory it is and the fact that, unlike the psychometric tradition, it is an adequate basis for a psychology of intelligence which could be used to provide an under-pinning for an educational system.

Has compensatory education failed?

Jensen asserts in his *Harvard Educational Review* article that compensatory education has been tried and has apparently failed. The reason he suggests is that black–white IQ differences are genetic in origin and so cannot be removed by education. By implication the poor whites are in the same boat as the blacks as they make up the majority of the recipients of the American intervention programmes. Most of this book has been devoted to an examination of these assertions but here we want to look more directly at the compensatory programmes themselves. Have they failed? Our discussion will be restricted to the American project Headstart, as British experience is too limited and our Educational Priority Areas are too recent to have produced any firm evidence.

Headstart was set up in 1965 as part of the War on Poverty. Its aims were not just educational but included the social, political and economic advancement of a sizable proportion of the country's inhabitants who were designated as poor. Achievement of the aims was not restricted to the provision of nursery schools but included a wide range of medical, social and community projects which hoped to involve 'maximum feasible participation of the poor' in their organization. So in our assessment we must look beyond the narrow educational goals.

The War on Poverty was launched in an atmosphere of optimism and enthusiasm which, at least in retrospect, seems to have prevented a careful analysis of what was being undertaken. The poor were to be helped so that they 'could take their place in the mainstream of American society'. But what would happen if they did just that? Are we not saying that they should become potential members of the middle class and, if they did that, would they not compete for the limited resources of housing, medical care, school places, jobs – wealth, power and prestige – that are available to the middle class? Suppose the programmes were successful, is it not likely that the middle class would object to sharing their limited resources and having them spread more thinly among a much larger number of people?

At least one Headstart programme does seem to have had sufficient impact to pose just this threat to the middle class and provoked them to stop it. These are the programmes of the Child Development Group of Mississippi which really did involve 'maximum feasible participation of the poor' and had dramatic effects on some of the poorest black communities in the United States. Anyone who saw these programmes in operation could not but be impressed by the way they had given pride and optimism to people who had suffered from generations of apathy and hopelessness. Careful assessment studies gave evidence of the improvements in the children's health, nutrition and social outlook. Their parents broke loose from the Sambo image of the share cropper and began to take an interest in their social and political environment. Poor blacks ran for and were elected to positions like mayor, sheriff and school-board membership – jobs that previously had been exclusively occupied by the white middle class. This led to an eventually successful campaign by the State Democratic politicians to have federal funding of the programmes cut off. The Headstart programmes were replaced by others organized and run by the white middle class.

So maybe we should conclude that Headstart had to be unsuccessful because success would have led to fundamental social and political change that the non-poor were not prepared to accept. It is one thing for politicians to use phrases like 'allowing the underprivileged to take their place in the mainstream of society' but quite another to translate such clichés into programmes of social action and to assess their long-term effects. All too often the jump from cliché to setting up of a pre-school programme was so rapid that no one paused to think out what was being undertaken.

Those who are now pronouncing compensatory education to be a failure use a single criterion – IQ. Headstart has failed, they say, because it has not raised IQ. We would concur with their evidence – some IQ gains have been produced but they have usually been short-lived and have often disappeared after a year in grade school. But enough has already been said about IQ to suggest that it is a highly misleading and limited measure to use, and as Joanna Ryan comments, there is some irony in the fact that the very people who have argued that IQ is a fixed attribute should try to use it as a measure of intellectual change.

For the purpose of our argument let us argue that Headstart may not have produced striking intellectual change; does this mean that all compensatory programmes are doomed? Surely we would have to try all possible programmes before that most improbable conclusion was justified. In fact, only a rather narrow range has been tried over relatively brief periods. Furthermore, we have good reasons for thinking that the basic assumptions of many of them have been too simple to provide effective strategies.

Most intervention programmes have been founded on the belief that 'deprived' children suffer from a language 'deficit'. Assuming that cognitive development and so school achievement are related to language and noting that poor children did not speak standard English (or American), the main thrust of pre-schools has been towards improving language skills. So, it was argued, if we want the poor child to behave and progress like the middle-class child we must teach him to speak properly. The difficulty here is in the statement that cognitive development is *related* to language. What does *related* mean? Most psychologists would accept that there is some connection but nobody, without venturing far beyond the current research, could specify what the connection is. The field is extremely complex and, until we have a little more understanding, we will not be in a position to devise a research-based intervention programme.

Within this broad picture there have been some local successes, and it is rather striking that these have occurred in programmes that have been based on an adequate notion of what may be involved in cognitive development. One would like to see an expansion of this sort of attempt where continuing research goes hand in hand with practical experiment and the whole thing is carried out with a realistic appreciation of the social context.

But, as may become clearer in the next section, it is not a straightforward matter to decide what kinds of cognitive change we should be aiming for. Attempts to boost IQ may not only fail to achieve their stated goal but also may be trying to obtain something we ought to reject if we could achieve it. Education based on the cult of the IQ score may be no education at all and amount to little more than a paternalistic display of middle-class values and prejudices. Therefore as a first priority we might do much better to seek to give *all* children the basic minimum for reasonable development – adequate diet, adequate housing and varied and pleasant environments for play. If these are not provided it is a misdirection of limited resources to provide elaborate teaching programmes especially where they are based on dubious psychological theories or no theories at all.

As Steven Rose has mentioned, there is hard experimental evidence that in animals nutrition is crucial in cognitive development. For humans the evidence is much less direct, but the overall picture seems sufficiently convincing for us to suggest that improving nutrition should be a major priority if we wish to foster cognitive development among poor children, even in countries apparently as well fed as Britain and the United States.

Adult height, perinatal mortality rate (babies dying in the period around birth) and life expectancy all provide indirect measures of the long-term effects of nutrition and medical care. Indeed they may be seen as indices of the health of a community. Within cultural groups, taller adults, coming from communities with low perinatal mortality rates and long life expectancies, tend to have higher IQs and to obtain greater educational success.[1] In Britain upward social mobility is related to adult height as well as IQ. There seems to be a vicious cycle from which the upwardly mobile poor have to break loose (see figure 6). These effects are transgenerational; perpetuated through the permanent damaging of the mother's ability to produce healthy babies and her family's lack of resources necessary for normal physical – let alone social – development. It is the healthier, taller children who escape from this environment and are upwardly mobile.

The picture in the United States seems very similar. A recent study by Doctors Breslow and Klein at the University of California compares the school achievement and health of various ethnic groups in California. The figures speak for themselves (see figure 7).

For each ethnic group, a whole range of health measures seemed to move hand in hand with the school achievement data. The authors suggest that 'there may be some underlying factor(s), e.g. some aspect of family or community life, responsible for both better health and more education'. Simple factors like frequency of visits to a doctor or income (whites

1 In this part of our discussion we are forced to use IQ as the measure of intellectual development as this has been the only measure used in the research. However, the effects of malnutrition are very generalized and there is every reason to believe that they would influence most aspects of intellectual development.

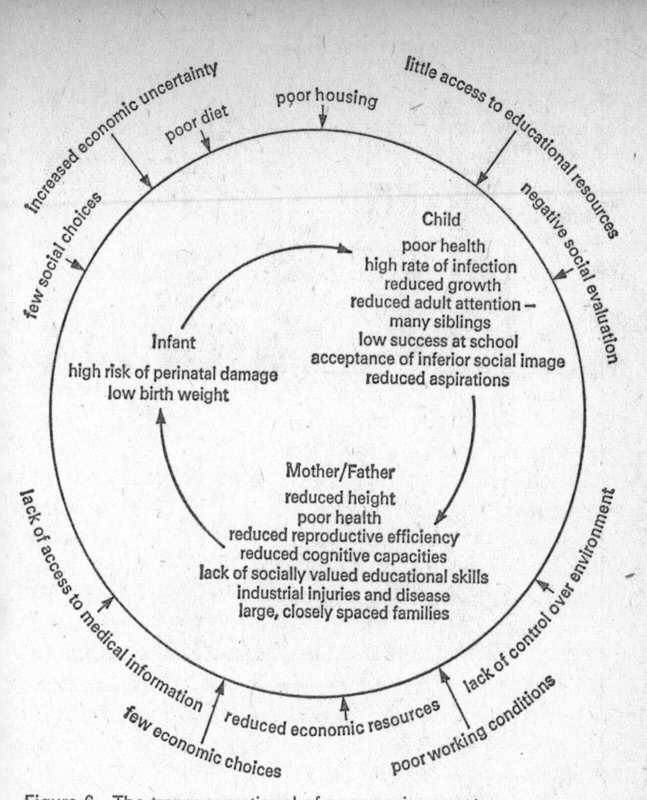

Figure 6 The transgenerational of poor environment

earn more than Japanese) cannot explain the relationship, but a combination of diet and social context is at least consistent with the findings.

Certainly, there has been some improvement in the general picture; children are growing into taller adults and IQ norms are rising. However, there is little evidence that the differentials within our societies are reduced – indeed the reverse might be true – so that, for example, while national perinatal mortality rates decline, the middle-class–working-class or black–white differences are static or increasing.

So to conclude our argument about the results of intervention programmes; to simply say they have failed is highly misleading.

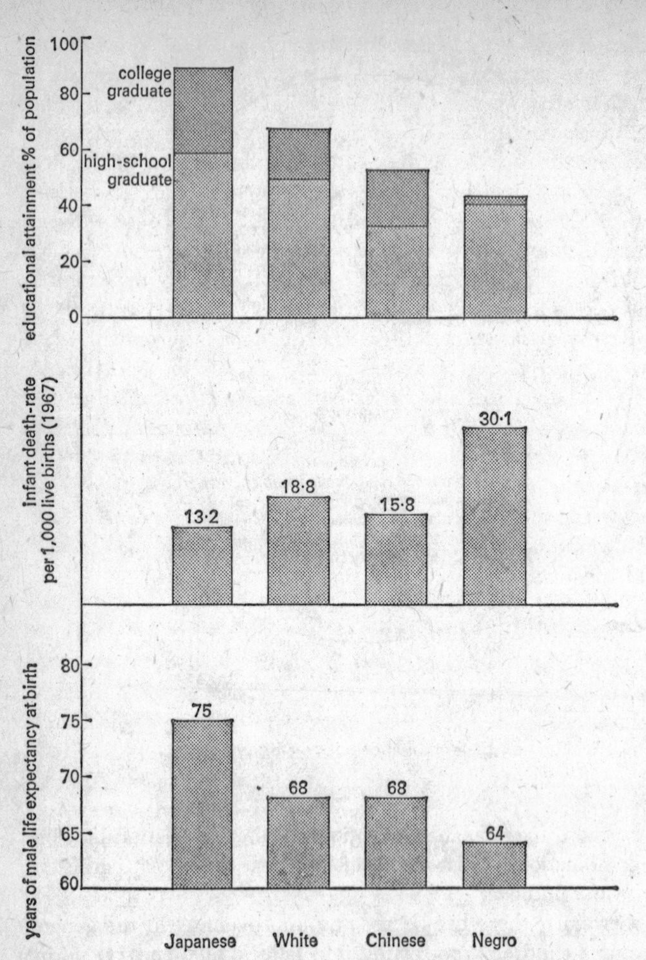

Figure 7 Community health related to educational achievement in California.

If we take the traditional assessment by means of changes in psychometric test scores, it is true that only rather limited effects have been demonstrated. But we ought to have severe reservations about whether this is the sort of intellectual change we ought to aim for. Without considering changes in IQ, there are other criteria we can use for judgement. Some programmes may have succeeded because they produced improvement in children's health and nutrition – this was particularly striking in Appalachia and the Deep South where it was only through Headstart that some children got any medical care, and for some of them it provided a substantial proportion of the daily food intake. Or we could justify nurseries simply because children enjoyed going to them and they freed parents from 24-hour child-minding. 'Sesame Street' may not raise IQ but children may like watching it.

Finally, we have seen there are fundamental problems when the political clichés approach reality. To lift the poor underclass and to give them equal opportunity to compete in the 'mainstream' entails drastic social change throughout society. The indications are that this was not foreseen or intended when the war on poverty was launched and that majority groups are not willing to accept such change. This raises the whole question of the relationship of educational innovation and social change which we will discuss in the next section.

Educational innovation and social change

At this point we would like to widen the debate a little further and consider some of the broader questions about educational innovation. As we have already mentioned, the traditional view of compensatory education programmes is that poor or 'deprived' children lack something which, if replaced, will allow them to participate in the mainstream of society. This view seems to involved an underlying assumption that educational innovation is a means of effecting social change – it is through education that the poor will get jobs and so break the cycle of poverty. Already in our discussion of the work of the Child Development Group of Mississippi we saw that if educational (broadly defined) programmes begin to influence social organization and threaten the *status quo*, they may be resisted. This should awaken us to the possibility that the educational system reflects the social structure and that changes in social structure may lead to modifications of the educational system, but that the reverse may be no more than a pious hope.

Psychological work tends to take a rather simple-minded

model of the educational system, assuming that within broad economic constraints schooling is provided for children in accordance with their abilities to profit by it. And that beyond school, selection for different occupational roles is based on combination of ability, which is taken to be synonomous with IQ, and the extent of specific training and skills appropriate to the job.

However, we could take a very different view (as was done in chapter 4) and see educational systems primarily as means for perpetuating social structures by allocating roles with their attendant power and resources – or lack of them – and by passing on the values inherent in the social structure.

Both Britain and the United States have pyramidal social structures with most of the power and resources concentrated among a small minority at the top of the pyramid. Social mobility is, in theory, quite unrestricted but, in practice, a high proportion of children end up in the same position as their parents. It is commonly supposed that social mobility is high but this is a false impression created by an overall change in the shape of the social pyramid, so that it has become less steeply tapered and almost diamond shaped during the last few decades. Or to put it in other words, skilled and white-collar jobs have increased as a result of increasing use of technology in industry, and so there has been a net upward migration. However, in proportional terms little has changed and, for example, the percentage of working-class children at British universities is just about what it was in the 1930s, though of course there has been an increase in the total number of students in higher education.

The pyramid of the social structure is echoed by the pyramid of the selective educational system and the shapes of the two pyramids change in parallel, with the social system slightly leading. Many examples can be provided which support this kind of connection. Changes in education are brought about to meet social and political needs rather than for educational reasons. The expansion of scientific and technical education following the launching of the first Sputnik is an example of one such change. On the other hand if we look for examples where educational innovation has been tried as a means of social change we see a succession of failures. The postwar reorganization of the British schools is an excellent example here – 11+ selection and then comprehensive schools were both attempts at providing equality of opportunity for all sections of society, but the effects on social mobility and distribution of resources in society seems to have been minimal.

Just as status and prestige are ordered hierarchically in society so the selective education system is organized along a dimension of ability. In education there is much talk of treating children as individuals but all too often this is lip service and judgements are made on the basis of a relative ranking of ability. Ability here is IQ or a notion very closely related to it. Such a unidimensional view of children certainly does not correspond with our day-to-day experience of individual people, which suggests infinite variety of kind, quality and style. Nor does it fit with our theories of cognitive and social development which do not contain single ordered hierarchies. It is possible to use Piaget's theory of intelligence to derive measures of the stage of cognitive development a child has reached and compare this with IQ or school progress. When this is done very small or zero correlations are found. This indicates that, at least within a 'normal' range, cognitive development has very little to do with one's success at school.

Michael Young, and more recently Roger Herrnstein, have suggested we are approaching or have reached a meritocracy situation where status and income are distributed solely on the basis of IQ. Assuming a high heritability of IQ, they argue that this will fix and accentuate genetic difference between social classes. John Hambley has already mentioned some of the genetic fallacies in this hypothesis. Here we should note that the the correlation between occupation and IQ is in fact quite low. Certainly we would not expect someone with an IQ of 70 to become a doctor or executive civil servant but a manual worker obtaining a test score of 120 would not be exceptional. Furthermore, if one just took the majority of people with IQs of around 100 the correlation would probably disappear and social class of parents would become a much better predictor of children's occupation and earnings. Education and society do both have unidimensional hierarchies but this need not imply that there will be close correlations of IQ and status.

Given the present hierarchical social system, there is no reason why a selective educational system should not provide equal opportunities to all social groups, provided that selection did not involve criteria that were correlated with social origin or the degree to which children accepted the social and political values of those at the top of the pyramid. But we have good evidence that selection involves just that. This argument is put very well by Bernard Williams in his important essay on equality:

Suppose that in a certain society great prestige is attached to membership of a warrior class, the duties of which require great physical strength. This class has in the past been recruited from certain wealthy families only; but egalitarian reformers achieve a change in the rules, by which warriors are recruited from all sections of the society on the results of a suitable competition. The effect of this, however, is that the wealthy families still provide virtually all the warriors, because the rest of the populace is so undernourished by reason of poverty that their physical strength is inferior to that of the wealthy and well nourished. The reformers protest that equality of opportunity has not really been achieved; the wealthy reply that in fact it has, and that the poor now have the opportunity of becoming warriors – it is just bad luck that their characteristics are such that they do not pass the test. 'We are not,' they might say, 'excluding anyone *for* being poor; we exclude people for being weak, and it is unfortunate that those who are poor are also weak.'

This answer would seem to most people feeble, and even cynical ... one knows that there is a causal connection between being poor and being undernourished and being physically weak. One supposes further that something could be done – subject to whatever economic conditions obtain in the imagined society – to alter the distribution of wealth. All this being so, the appeal by the wealthy to the 'bad luck' of the poor must appear as disingenuous.

It seems then that a system of allocation will fall short of equality of opportunity if the allocation of the good in question in fact works out unequally or disproportionately between different sections of society, if the unsuccessful sections are under a disadvantage which could be removed by further reform or social action. This was very clear in the imaginary example that was given, because the causal connections involved are simple and well known. In actual fact, however, the situations of this type that arise are more complicated, and it is easier to overlook the causal connections involved. This is particularly so in the case of educational selection, where such slippery concepts as 'intellectual ability' are involved. It is a known fact that the system of selection for grammar schools by the '11+' examination favours children in direct proportion to their social class, the children of professional homes having proportionately greater success than those from working class homes. We have every reason to suppose that these results are the product, in good part, of environmental factors; and we further know that imaginative social reform, both of the primary educational system and of living conditions, would favourably affect those environmental factors. In these circumstances, this system of education falls short of equality of opportunity.

IQ tests or measures closely correlated with them are widely used in educational selection and we have seen that these, far from being objective measures of intelligence, incorporate many values highly correlated with social origin. Over and above this, even when one compares children of equal IQ, we have clear evidence, as from Douglas's studies, that educational success is more often given to the children from the higher social class. Minority group children seem to be in much the same position as white working-class children.

If this view has any validity it means that in the long run the role of intervention programmes as instruments for social and economic advancement of minority groups or lower social classes is minimal. It does not follow that we should not have such intervention programmes as we have seen that they may have many other justifications, but we should not expect them to have any marked effects on poverty or the equality of educational opportunity. Looking at the narrow, 'cognitive', aims of such programmes, even if successful, they will not give equal opportunity to the poor because intelligence, in its true sense, is not the sole or major determinant of success in the educational system.

This discussion of the relationship between social structure and education has had to be very brief and is probably oversimplified. However, it does indicate that we should not expect too much from psychology as a means of modifying social organization. Its role is less central and more indirect than some in the trade have asserted. The psychologist's role is further delimited by the distinction between scientific assertions and moral, political and ethical judgements. The crucial issue is the allocation of resources, and psychology does not hold any particular brief to tell us what system of allocation should be operated, though it may play a part in operating a system once it has been chosen.

Eysenck, particularly in his Black Paper article, has attempted to provide a psychometric justification for the *status quo*. He can give the superficial impression of having a persuasive case because he represents such concepts as IQ as imperfect but objective and value free. He ignores the kinds of evidence presented in this book which indicate that the inherent values of such concepts reflect the social system from which they are derived. Such a brand of psychology is bound to support and make seem inevitable the system from which its values are taken. Claims for objectivity must be based on a careful analysis of the covert assumptions in concepts as well as on obedience to certain rules of argument and logic. Joanna Ryan and John Rex, beginning from very different standpoints, have sketched some of the analysis of assumptions that is called for – an analysis we find totally absent in psychometrics.

It seems an open question how far this process can be carried; we do not know whether a value-free psychology is *in principle* possible, and many doubt it. Nevertheless, we must strive to make our assumptions as open as possible. This requirement is all the more important when scientists are writing for the general public.

Even when the analysis of assumptions has been carried as far as possible, it will not result in a 'scientific truth' which is prescriptive for the solution of social problems. Such answers lie in a realm which is quite distinct from that of psychology. This is because they involve questions of evaluation of human worth and the division of goods between individuals. There are ethical, political and moral judgements which have no place in science, at least as currently conceived. Perhaps this is clear from the example of capital punishment. Science can certainly help to decide whether criminal behaviour is related to the punishments given to offenders but it could not tell whether it is right to kill. Socially responsible science must be based on this distinction. Even if psychologists were able to demonstrate the unlikely phenomenon that a socially defined group of people had quite different cognitive capacities from the rest of us, this could not on its own provide a reason for treating them in a special way. That decision rests with the whole of society.

Further reading

On the vicious circle of poverty, ill health and lack of education

L. Breslow and B. Klein, 'Health and race in California', *American Journal of Public Health*, vol. 61, pp. 763–75, 1971.

R. Illsley, 'The sociological study of reproduction and its outcome', in S. A. Richardson and A. F. Guttmacher (eds.), *Childbearing*, Williams & Wilkins, 1967.

On Piaget's work

J. H. Flavell, *The Developmental Psychology of Jean Piaget*, Van Nostrand, 1963.

J. Piaget, *The Origin of Intelligence in Children*, Norton, 1962.

J. Piaget and B. Inhelder, *The Psychology of the Child*, Routledge & Kegan Paul, 1969.

Cognitive psychology and education

J. S. Bruner, *Toward a Theory of Instruction*, Norton, 1968.

S. Farnham-Diggory. 'Cognitive synthesis in Negro and white children', *Monograph of the Society for Research in Child Development*, vol. 35, no. 2, 1970.

A. R. Jensen, 'How much can we boost IQ and scholastic achievement?', *Harvard Educational Review*, vol. 39, no. 1, 1969.

L. Kohlberg, 'Early education: a cognitive-developmental view', *Child Development*, vol. 39, pp. 1013–62, 1968.

On meritocracy

M. Young, *The Rise of the Meritocracy*, Penguin, 1961.

H. J. Eysenck, 'The rise of the mediocracy', Black Paper II, The Critical Quarterly Society, 1970.

R. Herrnstein, 'IQ', *Atlantic Monthly*, September 1971.

On equality

B. Williams, 'The idea of equality', in P. Laslett and W. G. Runciman (eds.), *Philosophy, Politics and Society*, Blackwell, 1969.

On development

W. Mary Woodward, *The Development of Behaviour*, Penguin, 1971.

Index